PLANNING AND MONITORING DESIGN WORK

E. J. COLES
C. M. H. BARRITT

Routledge
Taylor & Francis Group

LONDON AND NEW YORK

The CHARTERED
INSTITUTE OF
BUILDING

First published 2000 by Pearson Education Limited

Co-published with The Chartered Institute of Building through
Englemere Limited

Published 2014 by Routledge
2 Park Square, Milton Park, Abingdon, Oxon OX14 4RN
711 Third Avenue, New York, NY 10017, USA

Routledge is an imprint of the Taylor & Francis Group, an informa business

ISBN 13: 978-0-582-32029-1 (pbk)

British Library Cataloguing-in-Publication Data
A catalogue record for this book is available from the British Library

Set by 35 in (10/12pt EhrardtMT)

CONTENTS

LIST OF FIGURES

PREFACE

Until quite recently, building design and construction were invariably carried out as separate operations, but pressures for efficiency have been breaking down the barriers and brought new procurement methods such as design and build, design and construct and fast-track construction management. These options bring new project organisation structures and a migration of people into new roles.

Many books have been written about the planning of construction operations, but few note that design work must be managed in a different way. This book has been written to fill this gap and it is targeted at a wide readership, as follows:

Construction managers who wish to improve the quality of design and the flow of information for construction are likely to need specialised advice in this field.

Design partners and practice managers need work planning skills, to assist their staff to deliver clear, consistent and complete design information within project time-scales and the man-hour budgets allowed by design fees.

Clients need to understand the importance of supplying information and making decisions at the right time. This book might also help them to better appreciate what can be purchased for design and project management fees.

Project managers must understand how design professionals work together, so that deliverables are correctly defined, resources properly allocated and significant output recognised as work progresses.

Specialist sub-contractors carry out vital design tasks, often within very difficult time constraints. In order to control this work, they must ensure that reliable information is secured from the site and other designers. It may be much to their advantage if they know about factors that can constrain the availability of such information and how their own work could best be slotted into project programmes.

Lecturers in design and construction disciplines, who work with undergraduates, graduates and professionals in practice, should have a broad appreciation of the special characteristics of design work, how it is organised and how output should be fed through to site operations. They should also be aware of how the techniques used to control production are used differently for design work and construction operations.

Design students need to be aware of how interdisciplinary design teams can be organised systematically to achieve their objectives.

Construction students will wish to be aware of difficulties that can attend the control of design output, since the success of their future work may depend on this.

ACKNOWLEDGEMENTS

The thinking behind this book originated in the instruction and example of many colleagues in my early career, which I spent mainly in architects' offices. Attendance at the MSc course in Construction Management, at Brunel University, considerably broadened my appreciation of that experience.

Colin Gray was a major influence on my thinking, in the 1980s, when I worked for him as a Research Assistant in the Department of Construction Management and Engineering at the University of Reading.

The post at Reading enabled me to visit the Department of Architecture, Building and Planning of Eindhoven Technical University, in The Netherlands, where I enjoyed stimulating discussions with staff and postgraduate students. I am grateful for their gifts of publications, in particular, a copy of Jan Boekholt's PhD thesis, which has contributed to my view of design work.

I was diverted from Colin Gray's study of design information transfer to arrange a series of short courses for design managers, where I learned much from the presenters and the practising managers who attended. Colin also helped me to secure a Dennis Neale Scholarship, from The Chartered Institute of Building (CIOB), which allowed me to explore some problems of design management in practice. The continued encouragement of the CIOB Information Manager, Peter Harlow, has sustained me towards completing this book.

I think it is fair to mention the good work done by the organisers and presenters of the many seminars and conferences that I have attended, the most relevant of which were organised by the Royal Institute of British Architects, and the APM (now the Association of Project Management).

A number of people in construction disciplines have given their time to comment on the text, at various stages in this book's development. I was inspired to change so much that they may not recognise the finished product. In particular, my thanks are due to Ralph Levine, Graham Furness, Juan Callister and Ray McLean.

I was writing exclusively during evenings, weekends and holidays, which might otherwise have been spent with my family. I hope readers may share in my gratitude to them for accepting this with good grace.

E. J. Coles
West Lothian, August 1999.

INTRODUCTION

1.1 EFFICIENCY IN DESIGN

Clients of the building industry have been questioning its efficiency for many years. If substantial economies are to be achieved during construction and through the whole life of a building, expert management of the design work is needed. This is not only to ensure that the designs are cheap to construct, but also to put accurate and complete information in the builders' hands when they need it, so that construction can proceed smoothly and logically.

Building design is fundamental to civilisation. It expresses the concern of a culture for its institutions and the living and working conditions of its population. Good design is, of course, impossible without good designers. These people need the right conditions to be productive: they need good tools and materials, clear guidance about the project objectives, encouragement, recognition of their skills and above all, enough time. Without such support, the construction industry will neither attract nor keep the talented designers it needs and we will all have to live with the consequent effects on our built environment.

1.2 A DIVIDED INDUSTRY

Research into the efficiency of building construction has often highlighted the separation of designers from the construction site, which is considered to be a unique handicap of the building industry. By comparison, designers in the manufacturing industries work closely with engineers on the factory floor to ensure that their designs are efficient and that there will be a smooth flow of production. As a team, these people are also closely in touch with what is called 'the supply chain', to ensure that there will be a flow of suitable materials and components for the machinery of production lines and the products that come from them.

The rift between design and production in the construction industry developed in the UK over a long period of time. It originally came about because upper-class patrons did not wish to deal directly with builders, whom they considered were of the trading class. Architects and surveyors bridged the gap by designing in ever

1

greater detail, for approval by their patrons, before instructing builders to carry out the construction work. During the twentieth century, patronage for buildings has almost entirely moved away from the wealthy families and most work is now commissioned by commercial enterprises and government agencies. In the same period, building methods and services have become more complicated and consequently, the expert professional 'class' is still needed to design buildings and manage their construction.

The distinct classes may have disappeared, but there are still substantial differences between the sub-cultures of design and building companies. In the 1960s and 1970s, professionals still avoided competing with one another in the prices of their services, whereas building contracts were nearly always awarded to the company that promised the lowest price. Design professionals apply knowledge and creativity to develop abstract ideas in the form of drawings, calculations and specifications. Construction professionals juggle financial risks as they handle large quantities of material and manage expensive plant. Designers may need to persuade their white-collar clients, the public and authorities that design proposals will satisfy requirements. Construction professionals have to push projects through a challenging series of deadlines, using exceptional interpersonal skills to control the output of a succession of suppliers and an ever-changing population of gangs on site.

Within the design professions, the distinctions between architects, engineers and surveyors have become more sharply defined and differentiated over the years. This met the need for specialisation, as the technical complexities of construction grew. It also became a way for companies and professions to secure particular types of work and make it difficult for people without specialised knowledge to compete with them. In addition, the growing trend to litigation has encouraged the design professions to accept exclusion from areas where their competence may be limited. For example, many architects have withdrawn from supervising work on site and, as a consequence, they have less opportunity to learn about the real problems that are encountered by builders in using design information.

In response to the fragmentation of construction into design and other specialised areas, clients have sought ways to secure the smooth progress of work. Two of the more successful methods to emerge have been design and build and project management. In design and build, the client is protected from situations where a builder and the designers may each blame the other for faults in the building. Project managers minimise risks to clients by bringing expert management to design, construction and matters such as contract administration. Attention is also turning to the concept of project partnering as a means of encouraging participants to work together rather than concentrating, exclusively, on their own specialism.

1.3   PLANNING AND MONITORING TO REINTEGRATE THE CONSTRUCTION INDUSTRY

Construction projects comprise distinct phases of initial research to define requirements, design work, construction and commissioning (of building services and facility management). Many different companies and groups of people contribute to projects, especially during design and construction. To overcome the problems and inefficiencies of this fragmentation, the work has to be planned in such a way that each group has the information (and contracts) that it needs to start and will pass on necessary information to other groups at the time when their contribution is needed. Once such plans have been made, the progress of work has to be carefully monitored (observed) to ensure that everyone who contributes to the process is able to start and finish at the required times. Thus, planning and monitoring are key project management tools, whether work proceeds on a traditional basis, as design and build, partnering or on any other basis.

The Chartered Institute of Building sponsored a small-scale study of the factors that can prevent design processes achieving the required time, cost or quality standards.[1] The responses from the different professions suggested that the following factors were the most significant:

- Design authorisation by clients was slow or indecisive.
- Specialised design input came late.
- The logic of the design decisions was not effectively communicated, resulting in other designers and the construction companies failing to follow the intentions through correctly.

This list could be extended to show that every problem identified in this study was related to inadequate work planning for reasons such as:

- The knock-on effects of delay are not anticipated, that is, the dependencies of one task on work done in another task were not perceived.
- Resources were not deployed to their best effect.

1.4 THE NEED FOR PLANNING AND MONITORING OF DESIGN WORK

Design management in the construction industry is a live issue. To stay in business, most design practices have been forced to cut fees to the bone. Design departments in the public sector are also under pressure to show that they deliver value for money. This means that, if the quality of design output is to be maintained, if not improved, while fee income diminishes, the time and energy given to design work has to be very carefully controlled.

Many books have included sections about the planning of construction operations, but few have noted that design work should also be planned and monitored, or that this has to be done in a different way from that used to direct construction. Builders are given precise and detailed design information, which they analyse to derive construction activities. These activities can then be planned so that the building is assembled in a logical order, taking into account limiting factors such as site access, storage and cranage. Design work rarely has precise and detailed information to start from. Instead, it is the designers' job to create this. Until now, the methodology for planning and programming design work in advance has not been widely known. Nor have the difficulties of monitoring the progress of design work been taken into account.

It is of great concern that the effects of inefficiency in design procedures can be magnified during the construction phase and on, throughout the life of the building. Design information that is not clear, co-ordinated, sufficiently complete and correctly timed can lead to confusion on the site, substantial unplanned expenditure, unhappy builders and a dissatisfied client. If the design itself is inadequate, the building owners and occupiers will continue to pay dearly, through inconvenience and maintenance costs, for as long as the building exists. It is therefore of vital concern that the time spent by designers, which becomes less freely available as fees are reduced, should be used more and more effectively. The careful planning and monitoring of design work will assist greatly to achieve this.

1.5   DEFINITIONS

The term planning is used in many contexts, for example, town planners and regional authorities develop transport and land-use plans, while architects plan the physical layout of buildings and the land around them. It is therefore worth pointing out that, in this book, planning refers exclusively to the preparatory organisation of work, that is, the analysis of how people should co-operate in the use of time, skill, authority and resources to generate design information and use it to construct a building.

A programme of work is a list of tasks that have to be done, showing the times at which they should be done (generally calendar dates). This term can also be used in a wider sense, to include the logic that determines the best sequence in which the tasks should be done and the allocation of work (who should do what). Note also that, in this book, the term plan generally refers to a programme of work and not a drawing.

Monitoring, in this context, is the comparison of the actual progress of work against the planned programme. Substantial deviations from the plan indicate that action is needed, either to bring progress back into line, or to redraft an unrealistic programme and make it workable.

The term 'design discipline' is used extensively in this book. It permits generalisations to be made, without referring to a particular profession, or having to list them all, that is, architecture, building services engineering, structural design and the many specialist design services that exist. In general, the quantity surveying discipline is considered to be included, except where a sentence refers very strictly to *design* activities.

Similarly, the term 'consultant' is used in a general sense, referring to individuals or companies who are engaged in one or several design activities that need not be specified.

1.6   PLANNING AND RISK

The several parties engaged in a building project are exposed to different forms of risk and, as a result, their interest in work planning is likely to vary. The scope and detail of planning will also vary according to factors such as the type of project, its location, how the project is organised and the economic climate. The perception of risk by clients, project managers and design consultants may play a part in determining the amount of planning and monitoring that they decide to do and how much money they are prepared spend on it.

Prior awareness of the likely difficulties in beginning or completing tasks can guide the allocation of appropriate safety margins in the project programme and financial budget, to the points where they are most needed. Planning can also directly reduce risks, for example, the designers may not be aware of how tight the time parameters on their work really are, until they examine the necessary sequence of design tasks and the complexities of gaining approval for their output from the authorities and the client. If such difficulties are recognised in advance, work can be organised to minimise adverse effects, such as delay.

It should be appreciated that planning is not a 'big stick', which can be used to 'beat' people if they appear to be delaying a project. Such action would bring other sorts of risks, such as individuals and companies expending effort on protecting their interests, rather than getting the job done. Planning aims to increase co-operation, not decrease it. By illuminating the real problems that lie on the path ahead, these can be overcome and a rational view of priorities agreed across the design team and the project organisation.

One obstacle to the adoption of planning and monitoring techniques is uncertainty about their effectiveness. The expenditure of time and energy on these techniques may itself be perceived as a risk, especially where individuals in the design team have little experience in this domain. In particular, some managers may have adverse experience of attempting to plan and monitor design work as if it were construction. The purpose of this book is to reduce that area of uncertainty and bring to its readers a greater probability of success in managing design work.

1.7   PLANNING AND HASTE

If the period available for design work is short, careful planning can draw attention to the deadlines for key input, including decisions from clients. If such deadlines are not met, whole sequences of design activity can be delayed, with the consequence that work becomes rushed, many errors are made and delays occur during construction while these are sorted out. Planning helps to ensure that work is done in the right sequence and at the right time, with a minimum of reworking, so that no-one in a design team has to wait for the information they need to continue with their work. Planning should reduce haste and thereby allow time to improve the quality of designs and reduce the construction and running costs of a building.

In general, a shortened programme, or in other words, greater haste, requires more planning, not less. Where the time pressure is great, more attention must also be given to systematic monitoring to achieve the close control necessary to carry through the tight time plan. It should also be appreciated that a carefully considered overlap between the completion of design work and the start of construction operations can allow more time for both, rather than less.

Good work planning allows designers more time to consider how building components will be put together during construction. Success in this area depends on the competence of the designers and often on effective communication between them and the builders and suppliers. Where planning and monitoring are used in conjunction with good organisational skills and contract management practice, the time and money saved through good design for construction are likely to outweigh these management overheads many times over.

It is significant that the government has recognised that this organised approach may also improve safety on construction sites. Since the Construction (Design and Management) Regulations[2] became effective, the analysis of how buildings will be put up has become obligatory for all but very small projects.

1.8   BRIEFING AND DECISION MAKING

Any plan for design work must ensure that, where decisions are needed, relevant information is communicated to people who possess the necessary knowledge, experience and interest in the outcome to make well balanced judgements. This applies when clients intend work to be done by the designers, when specialist designers seek decisions from each other and when the designers seek decisions from their clients. It is also necessary that these individuals (or groups) are given authority from the clients to take decisions when these are needed, through some form of contract.

Any communication between the designers and someone without the same expertise or understanding should be carefully formulated for its purpose. Where decisions have to be made by non–experts, time should be given to explain all the relevant factors and for responses to be considered and discussed if necessary. A clear timetable of events will help to ensure that this happens smoothly and without unexpected delays.

1.9   LONG-LEAD SUPPLY ITEMS

Long-lead supply items are those skills, materials, components or assemblies which may take a long time to procure, either because they are in short supply or because many months are needed to design and manufacture them. Design information about these items may be needed well in advance of the date they are required on site, so that they can be ordered in good time. This generates critical information deadlines that occur before the majority of the design information needs to be finished. Such information requirements can have a significant effect on the sequence of the design work, the procurement of construction services and, possibly, on the design itself.

For example, if an external cladding system is proposed, its design may have to be integrated with the structural design and that of the environmental control mechanisms. To obtain early design input from the manufacturer of such cladding, it may be necessary to appoint the supplier in advance of placing other contracts. If this is not possible, then a different design solution might be needed.

Official approvals have some characteristics of long-lead items. For example, the time taken to secure building control approvals may require the early development and release of particular areas of the design information.

1.10 THE PARTIES TO PLANNING AND MONITORING

The companies and individuals engaged on a construction project all make different contributions and they are each affected in a different way and to a different extent by the processes of planning and monitoring. None, however, should ignore the planned programme or changes to it that may arise through subsequent progress; to do so would, inevitably, lead to the frustration of other members of the team.

1.10.1 The client

Private individuals represent a large proportion of the client population but many are companies, consortia or public bodies. Within these organisations are individuals who represent the client body and they will be accountable to their employers, counsellors or the government and, not least, to their financiers. One major concern for all clients, whether within an organisation or not, will be to ensure that the design programme allows sufficient time to agree the design requirements and proposals with everyone concerned. To establish just how long this time allowance should be calls for a detailed knowledge of such things as the approvals structure within the organisation, the degree of delegation permitted, the frequency of meetings of the approving committee and, not least, exactly who holds what authority. These factors will all affect the time to be allowed and must be built into the design work plan.

1.10.2 Partners in design practice

Professional institutions advise their members never to undertake a commission without arranging adequate resources to carry it out. The initial availability of skilled staff must be considered as well as the availability of business finance and the continuing allocation of these resources. This means that design companies have a professional obligation to plan and monitor their work with care, for they may be held legally liable if they are unable to complete their work satisfactorily.

Sufficient time can only be given to researching design options if the design practice or consortium is confident that enough time will be left, within the period allowed for the entire design process, to complete the design information for official approval and construction. Such confidence could come from long experience but, since most design projects are unique, early analysis and calculation of the programme of work would further reduce the risk of running out of time. Time also has to be set aside to check and co-ordinate the building design information at every stage of its development, so as to eliminate inconsistencies, gaps and inaccuracies.

Missing or confusing design information increases the time pressures in the building construction phase, which can, in principle, lead to counter-claims against the designers if construction work is delayed and damages are instituted. To meet this problem, design practice managers should endeavour to identify and agree the key dates when correct, complete and co-ordinated sections of the design information must be in the hands of the contractors. Having established these, the working priorities will follow.

Consultants within, or serving, design practices must not only analyse the design information requirements of contractors, suppliers and installers but also consider the time implications of preparing information for authorities, other consultants and any specialist designers with whom they must work.

A further use of work planning is to exploit developments in office technology. For example, to secure a rapid return on investment in computer aided design (CAD) equipment and training. The programmes of work for several projects can be integrated to ensure that the computers and the trained staff are used continuously. Although CAD is now an everyday tool, there may still be offices where bottlenecks in the work flow can be caused by the sharing of equipment, software or skilled operators. These matters can be further complicated in practice by the necessity to co-ordinate the work of consultants who work in different companies and who have different commitments and work load patterns.

1.10.3 Design team leaders

The function of a team leader in any design discipline is to develop design information within the time and economic constraints set by the fees that have been agreed for each commission. The team leader should endeavour to plan work so that the designers do not suffer the frustration that can follow from design decisions being taken in a poorly considered sequence, resulting in work having to be redone. Where many design projects are being channelled through one office at the same time, or in quick succession, it can be important to identify, in advance, the times when staff will be under the greatest pressure. Priorities, time-scales, and the general allocation of work can be re-examined if these 'pinch-points' are recognised in advance.

The designers in a team will often be heard to claim that their work has been delayed by others, who are slow to release necessary information. Such hold-ups can be avoided if the project programme is known and understood by the members of the other disciplines in the design team and they adhere to the timings set.

1.10.4 Designers

Creativity is often expected from designers. Its full value may only be realised if they have time to understand the problems they have been set in sufficient depth to perceive the options that are open, evaluate alternatives and develop the ideas that arise from such study. Planning the work flow will help to ensure that adequate time is available for these activities.

The commitment of designers to meet deadlines for their work can be improved, if they are brought into the planning of their own work. Discussion of the programme can also make them aware of the difficulties they could create for others if they do not complete their part of the work by the agreed dates.

It is seldom that an exercise as complex as the designing of a building will be carried through without any problems arising. A carefully considered work plan should help the members of the design team to overcome the obstacles that invariably crop up, without necessarily frustrating the contributions of others.

1.10.5 Project managers

Project managers are not engaged to contribute directly to the design output or construction work, but they can exert a substantial influence over its progress. A competent project manager will ensure that he or she does not slow communication by becoming nothing more than a 'postbox' between the client and the design team.

A project manager's responsibilities may be defined more correctly as being to:

- Design the project organisation, including the allocation of responsibilities, the system of contracts and core communication patterns, for example to ensure that decisions are taken by the right people at the appropriate time.
- Procure the required services.
- Set-up and operate the management systems, including mechanisms for planning and monitoring the work.
- Provide leadership and drive.

Many project managers are drawn from non-design backgrounds. The procedure of developing detailed work schedules, with the leaders of the design team, can assist them to understand the design development process and agree realistic production targets. All project managers monitor output, in comparison with the agreed programme, to quickly recognise where their intervention is needed, for example, to give an extra 'push' to keep work up with the detailed targets, or to join the designers in revising their programme.

SUMMARY

The interest shown in planning and monitoring design work, by the various parties to a building project, roughly corresponds to the risks to which their work exposes them. In particular:

- Clients are anxious that their building is available for occupation at the planned date, which is often to accommodate a specific development of their business.
- Project managers take on their client's concerns to achieve the agreed delivery date for the building and use planning and monitoring to ensure that enough

time is given to exploring options and maintaining control over the quality of work.

- Designers aim to generate output of a consistently high quality yet, at the same time, preserve their profit margins.
- Contractors need to avoid the delays and the losses that can result from faulty or late design information.

The planning and monitoring of design work helps to satisfy these requirements by:

- Verifying that design information delivery targets can be met.
- Shoe-horning design activities into restricted fee budgets and time-scales.
- Increasing motivation by drawing attention to risks.
- Demonstrating the need for the timely input of information, including briefing and client decisions.
- Communicating who should do what, when and why.
- Pre-empting problems as work progresses, by raising the designers' awareness of one another's work and the procurement programme.
- Reducing haste, by building-in and protecting safety margins where they are most needed and thereby providing time for creativity and work of better quality.
- Integrating the work of different organisations, so as to bridge the 'cultural gap' between people who work in the various design disciplines, client organisations, statutory authorities and construction companies.

REFERENCES

1. Coles, E. J. 1990 Occasional Paper No. 40: Design Management – a study of practice in the building industry. The Chartered Insititute of Building.
2. The Construction (Design and Management) Regulations 1994, made under the Health and Safety at Work Act 1974, enacted January 1995 and enforcible from 31st March 1995.

CHAPTER 2

PROCESS CONTROL

2.1 THE PLACE OF PLANNING AND MONITORING IN PROCESS CONTROL

Production design and control techniques are used to ensure that output achieves the targets that have been set for time, quality and efficiency. A programme which is devised for building design work is, in fact, a design for a production process which also informs people about what they should be doing at any given moment. With this programme as a basis of comparison, managers monitor progress in order to recognise when output deviates from the plan and take remedial action. Work planning and monitoring can therefore be considered as production design and control techniques.

If the quality of output is unacceptable, work has to be redone. Thus, the value of planning depends entirely on the effectiveness of the control exercised over the quality of work. For many kinds of work, including design, the relationship is reciprocal, because the quality of work can suffer if insufficient time is allocated to tasks, or if a designer with appropriate skills is not available at the right time. Senior designers, who control the quality of the output of their team, therefore have an interest in ensuring that sufficient time is available for each task and that this time is used effectively. This book concentrates on the planning and monitoring of the time taken to do the work, but frequent reference is made to quality control, to follow through the implications of this mutual dependency.

Managing people is different from controlling a production line, where poor output can often to be rectified by correcting machine malfunctions and defects in the material that is being used. Nevertheless, the principles used to design and control industrial processes are, to some extent, applicable to design work. This chapter therefore outlines the basic concepts of process control, in order to make the following chapters easier to understand.

2.2 BASIC PROCESS CONTROL

Production processes in any industry can be analysed in terms of their input and output and how these link to and from the other production processes. Many aspects of input and output can be measured. These measurements are the key to

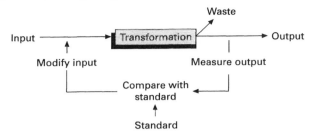

Fig. 2.1   A basic process control model

controlling the quality, cost and timely delivery of the product, because deviations from the standard for any of these aspects can be the cue for corrective action to be taken.

Figure 2.1 shows a general model for all kinds of work. The 'input' can include materials, settings for machinery, energy, skill and instructions to people. The 'transformation' could be a continuous process, a repeated process or a single task. The 'output' is the product. (These terms are often used in computing, but that is just one application of a principle that is relevant to a very broad range of products and services.) Waste inevitably forms part of the output, because no transformation is 100 per cent efficient.

The labels on the return arrows 'measure output', 'compare with standard' and 'modify input', outline a series of actions called a *control feedback loop*. The purpose of this loop is to alert operators, supervisors and managers to shortfalls in the quality or quantity of the output, so that this can be made good without delay and consequent loss.

Output may be measured in different ways and for different reasons. The quality of output is always important and the quality 'standard' can be quite a sophisticated test, or consist of a large number of tests. In building design work, each design discipline applies its own set of 'design criteria', with the client's brief as the primary, integrating standard.

2.3   ELEMENTAL DESIGN PROCESSES

Unlike industrial production, design work is largely unrepetitive. Nevertheless, it can be seen as a series of processes, which use the experience and imagination of designers to transform a brief into a clear and precise set of instructions from which construction companies can build. These instructions are, in effect, a detailed model of the building. This model often includes information in different forms: drawings on paper, electronic data, calculations, sample boards of materials and colours, and sometimes actual elements of construction which prove that the parts will fit together as intended. These models can be tested in many different ways, to be sure that every aspect of the design meets the client's requirements and will pass the scrutiny of the regulating authorities.

Designers generate these design models by carrying out a multitude of small tasks that may be referred to as elemental processes. Each of these tasks has the essential characteristics of the transformation process shown in Figure 2.1. In order to measure output and adjust inputs, for the wide range of design activities that must be carried out, many quality (and time) standards are needed. As design information has to be effectively communicated, in addition to representing the building as it should be, many of the control standards refer, not to the building, but to graphic and other communication conventions and protocols. For example, a builder was once puzzled by a window that did not fit the opening he had made for it. In this instance, the detail drawing had been drawn half full-size, but the joiner assumed it to be a full-size drawing (referred to as a rod in the trade) and had carefully made a half-size window!

2.4 ACCEPTANCE CRITERIA FOR OUTPUT

Building designs must measure up to a very large number of standards, including the building regulations, British Standards, many codes of practice, the requirements of the building type and, not least, the preferences of the client. These sets of standards can be referred to as acceptance criteria or compliance criteria.

For a very large part of the output of design work, the acceptance criteria are not set by the specialised staff producing the design. For example, the structural members of a building are often a conspicuous part of the image the architect is trying to realise. Similarly, the services engineer has to keep the pipes and ducts tidy, in a way that the architect will accept, and has to locate these elements so that they have secure fixings to the structure but don't actually go through essential structural elements such as the reinforcement in beams. There can be great difficulties if designers in one specialisation do not adequately understand the acceptance criteria of another.

To monitor progress accurately, the quality of output has to be measured and validated. It is not uncommon for designers to believe that they have completed a task, whereas they have been working to the wrong criteria and the work has to be done again. Extra rounds of design work can set work programmes back considerably, especially if key areas of the design, which matter to several design disciplines, are returned for further work. This can put everyone in the design team under pressure and can lead to further errors. In these days of narrow profit margins, a design company may have no staff available to correct the design information that is passed on for construction. The client and building users may have to live with the consequent inadequacies in their building.

This is a pivotal management consideration, since abortive work (as with any other kind of waste) eats into the profits and competitiveness of a design practice, or signals inefficiency in the building design departments of companies or government organisations.

A systematic approach to the management of quality is therefore indispensable. Work planning, itself, benefits if there is a structured methodology that can be planned along with that of design production. In this way, the generation and application of acceptance criteria can go in parallel with design development and avoid the rework that could otherwise invalidate the time plan for the work.

2.5   CHECKING WORK AS IT IS DONE

Drawings produced on computers can be transferred rapidly between the engineers, architect and other specialists, by electronic mail or diskettes sent by post. This allows additions and changes to the design to be checked by all members of the design team on a day-to-day basis. However, there is a big difference between the *potential* to integrate designs in this way and actually being able to do this in practice.

Some checking can be done automatically by a CAD programme, such as 'clash detection' which draws attention to any location in the model of the building where two designers are trying to position at least part of different physical items in the same place. Other levels of checking require alert attention and considerable imagination. For example, it can be quite hard to recognise from drawings that the noise from plumbing or plant could travel around the building structure and be a nuisance. This certainly won't show on the computer!

Time should be allowed for the design output of different designers to be compared. Inconsistencies can be found even in the output from the same designer, at different times. Designers may give low priority to checking their work, because they see this as a distraction from productive effort. It can also be difficult for an individual to see faults in designs that he or she has worked on. This is why formal design reviews are held, when productive work is temporarily set aside and the designers from different disciplines get together to pool their impressions, quite often in the company of a client representative and experts who have had no part in the production processes. On very large projects, it can be worthwhile to employ experienced individuals specifically to check design output.

In most design practices, output is not checked continuously, but efforts are made to check sets of related design information regularly. Experienced designers can focus on aspects of a design where inconsistencies between the work of different design professionals are most likely to occur.

2.6   DEPENDENT PROCESSES

Products and services are, invariably, the output of a series of interrelated processes. Figure 2.2 illustrates a simple fragment from what might be a very complicated network of interlinked production processes. It shows that, where production follows a series of processes, standards have to be established for each, or control is not complete. For example, if the output of Process 1 has a fault that is not detected, then a faulty input is fed into Process 3.

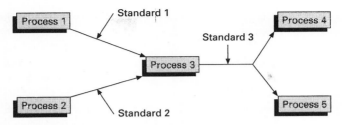

Fig. 2.2 Interdependent production processes

If an item goes through a series of production processes with an undetected fault, the work done on it will be wasted, until the fault is found and the product rejected or set aside for correction. If faults go undetected, then more material, energy, time and money is lost at each step.

If a fault remains uncorrected when a product or service is sold, the consequences can be serious. With complicated products such as design information, where it is very difficult to test that output is perfectly formed, faults are quite common. In practice, many faults in building design information are found during construction and some of these can be expensive to correct. Occasionally, faults come to light only after a building has been finished and occupied. In recognition of this risk, designers buy professional indemnity insurance to protect themselves from legal claims that can follow when design faults are found after the the building is handed-over to its owner.

2.7 TOOLS TO CONTROL QUALITY AND THE FLOW OF INPUTS AND OUTPUTS

Over many decades, building design professionals have developed systematic ways of working, in order to minimise faults in their output. Nowadays, the management of quality in design work is often formalised in a quality system. The system comprises a manual of quality control procedures and their systematic application. The procedures include checks that are made, on design input, on methods of working and on output, to ensure that work is properly executed.

Quality management procedures are applied in building design work at several levels, for example, the quality of the design itself, instructions to builders about the quality of construction and the flow of communication that supports design work. The following examples illustrate the variety of quality management tools and their application.

2.7.1 Quality control sheets

In essence, these are tick sheets against which the adequacy of preparation for a task, the steps of the process and the acceptability of its output, can be checked. Quality control sheets must be prepared in advance of the work. Sufficient time has therefore to be allowed before work begins, to check that standard quality control sheets are applicable to the particular work in hand and generate new ones if necessary.

The use of quality control sheets in practice is limited by several factors. They take time to prepare and it is unlikely that every pertinent factor could be documented in advance. They also take time to use, and this may seem an unnecessary encumbrance, when professional designers carry many of the acceptance criteria in their heads. It may be necessary to disregard some factors if it turns out that some of the acceptance criteria are incompatible, or cannot be achieved within the time allowed for design development.

It is always important to keep the overall aims of the project in view and appreciate that outcomes, as measured by the client's policies and strategy, may override the more detailed application of quality control. There is, in fact, a hierarchy of acceptance criteria that has to be respected.

2.7.2 Specifications

Design information for construction invariably includes extensive documentation, which sets out precise acceptance criteria for almost every material, component and assembly which is to be included in the building. The specification may also include extensive guidance about how construction operations are to proceed, including the precautions that are to be taken to protect neighbours of the site and the public from nuisance. The specifications are developed in parallel with the design drawings, because the need to communicate specific acceptance criteria to the builder is usually recognised as the design work progresses. Sometimes, it is a good idea, if not essential, that these criteria should be set out in advance of some areas of the design work, to tighten control of the design development.

2.7.3 Document management systems

Such systems play a vital part in controlling the development of complex design work. A document management system may be used throughout design and construction. Its primary purpose during design is to ensure that every designer bases their work on correct and current information. Before the advent of computing, document issue sheets were used to record what information had been passed between the design disciplines and to the quantity surveyors, the client and the authorities. Developments in information technology now permit records to be kept on computer and this database may be viewed on a display unit by any member of the design team, through the telephone network.

The development of electronic information technology has also allowed designs in development to be seen directly by designers in other offices. However, this does not make it easier to keep track of the precise status of information as it is developed or to ensure that every member of the team is using current and correct data. Some practices and projects employ information managers, who, among other duties, will routinely ensure that the design information in use is up to date and co-ordinated.

In the absence of an adequate document management system, faults can develop in design output, which may undermine the forward planning of work. Additionally, the document management system can give managers a valuable view of progress, by which to measure whether or not the work is proceeding according to its programme.

2.7.4   Design co-ordination meetings

Even with sophisticated information technology, it is not always possible for designers to keep sufficiently in touch with the work of other specialised designers in different offices. This is because matters such as design strategy and acceptance criteria are not generally accessible through the computing systems. Regular design co-ordination meetings are therefore needed to discuss current work and ensure that every member of the design team knows where they should take account of designs being developed by the others. Teleconferencing facilities can be used for these meetings, where the various participants have access to precisely the same drawings and other documents, for example, through networked computers.

2.8   ACCEPTANCE CRITERIA FOR INPUT

Referring again to Figure 2.1 (page 12), if the output of any process is awaited before any adjustment is made to the input, it will be too late to exercise control over the process. Quite simply, checking output for faults is putting the cart before the horse. What comes out of a process is much more likely to be acceptable if careful attention is paid to what goes into it in the first place. Attention should also be paid to the working methods that are applied to develop the design, as these are part of the input.

Communications should be checked for accuracy, because these are frequently direct or indirect inputs to design work, either as partly finished designs, passed to another designer, or setting out the acceptance criteria and working methods. These communications include those with the client, authorities and other bodies normally considered to be outside the design process. If all these inputs are attended to with care, the chances of the design output being acceptable will be greatly enhanced.

The project definition and brief are key inputs at the beginning of a design process. Appropriate acceptance criteria for all areas of a design grow out of these statements. Producing a good brief, and communicating it effectively to each section of a design team, is a task that requires skill and experience. If this is not done properly, input and acceptance criteria are likely to contain faults; muddle will follow and, however well the work programme might be planned, it is unlikely that the design team would be able to keep to it.

The design team itself is a key input to the work process. Architectural designers and other consultants have to be selected with care. There should be people in each part of the team who thoroughly understand the building type and the procurement method to be used for construction. In fact, there are many criteria for selection, but these are the subject of employment and personnel techniques that are outside the scope of this book.

No-one would use a machine for production operations in a factory without first checking its calibration. Selecting the designers could be considered analogous to choosing the production machine, in which case it can be seen that, to calibrate their performance requires examination of their present views of the building type and, possibly, those of the client and anticipated users of the building. New members of the design team are likely still to be conditioned by previous work, and, consequently, possessed of an inappropriate mind-set. It is also possible that some of them may not have worked on a project that is similar in every respect, so their minds need to be cleared of any preconceptions they may have and introduced to the realities of the project on which they are starting. Carrying out an induction appropriate to the many different professions and individuals, probably at different stages of the design process, is also a skilled task, requiring considerable experience of the building type, client, procurement method and working in multi-discipline design teams.

The mind-set of designers can be harder to control if individuals are working on several design projects at the same time. However, on smaller projects, the tasks are usually relatively simple, and this is not necessarily an impossible exercise for those involved.

Creativity is an input that defies measurement, although it will no doubt be sought when the designers are selected for a project. It is an aspect of a project organisation that is difficult to define in the brief or appointments, but most clients will be looking for people who have created designs before that empathise in some way with the sub-culture to which they belong.

A particularly important input is the authority to proceed. Without this one section of the design team may turn to other work, while waiting for approved output from another discipline, for fear of wasting the limited amount of time allocated to the project.

Besides these important inputs, there are everyday inputs that are easily taken for granted. For example, designers have to turn up for work and stick at it for the necessary time. There is the transfer of information between designers, that can so easily be misinterpreted if it is incomplete or not well formatted. There is a plethora of published standards, guidance notes, trade literature, regulations and so on. The key thing to grasp is that if any of these things is incorrect, inappropriate, incomplete or misinterpreted when it is brought into a design process, it will almost certainly have an effect that has to be corrected later on, together with whatever damage it may have done to any other parts of the design it has influenced. Weeding out faults, after they have been built-in to a design can be horrendously difficult. Unless rigidly controlled, they can be passed on to the construction phase and many take physical form in the building itself.

It cannot be overemphasised that skilful management of the inputs to design work is vital to achieving a high quality of design output. It is necessary for the efficiency of the process, its profitability and timely completion. If design input is not well managed, the reputation of every party to the project is at risk and some may be ready to point a finger of blame.

2.9   RANDOM INPUT AND 'NOISE'

It may at first seem paradoxical that fault tolerance is a cornerstone of secure quality management! No-one in the design team should be made to feel reluctant to disclose errors, or even suspicions that something might not be quite right, for otherwise, errors are likely to be compounded, to the detriment of the design and the programme of work.

Some unwanted input invariably creeps into design work. Bugs in computing systems, for example, are legendary and no-one may assume that their system will function perfectly and be safe from random or corrupted information. Interconnections between systems in different design offices increase the possibilities for losing or corrupting data.

One well recognised common error that can disrupt the smooth flow of design work occurs when someone continues to work on a superseded version of a design, so that the co-ordination between designs gradually disintegrates.

Assumptions are frequently the most dangerous 'random input', largely because people do not realise that they have been made. For example, it is very easy for designers to assume that, because they have been appointed to a particular project, the client is going to accept their every recommendation. Not all factors affecting the operation of a building will be apparent during the briefing process. These may come to light much later, for example when a client responds to a drawing or reads a specification and then realises that the designer has got something wrong. Sometimes, a designer may never come to understand what was unacceptable in his or her proposal, because there is simply a difference between their preference and that of a client.

It is not always appreciated that a client, or the representative of a client organisation, may not have acquired the specialised knowledge needed to read drawings and interpret the proposals shown to them. They may not have time to check documentation when they should and, as a consequence, parts of the building may actually be constructed before the inadequacy is recognised.

With many design tasks, one person has to hold all the relevant facts and options in their mind at the same time to generate appropriate solutions. In the event of sickness or a designer choosing to leave the practice, other designers can pick up the threads of the design problem from partially completed work, but this may cost time.

Although holidays are not generally taken at random, if work slips behind schedule, it could happen that key individuals are away at the very moment when their knowledge of the project is really needed. Probably everyone in business has experience of 'burning the midnight oil', just before taking a holiday, to be able to pass on a section of work to someone else who needs this while they are away. Work under such pressure is unlikely to be properly checked. It is a hallmark of a real expert to produce output of the required quality under such circumstances.

Thus, managers should not be overly critical of an individual if there are faults in work. There are bound to be faults. Probably the most important management skill is the ability to set up a quality culture that ensures full and fearless co-operation between the project partners, who draw attention to problems without delay and deal with them. If design work is to progress without hindrance, this attitude should also extend to the client and the construction team.

2.10   THE PROBLEMS OF ONE-OFF PRODUCTION

Consistent quality control is relatively easy to implement where products roll off a production line. In this situation, the quality checks on materials, production machinery and so forth can be planned in advance and refined through daily use over months if not years. Building design processes are not like this, because each design is developed only once. Although there are many similarities between the design processes for successive buildings, they never repeat precisely what was done before and quality management has to take these differences into account.

Standards for design input and output should be set carefully, or rework may be needed, leading to delay, loss of staff time and profit for the design team. For example, if the level of noise from an air-conditioning system is not taken into account at the right time the consequences may be significant. In this instance, the client's representative might have been given the predictions of the noise emission level, without appreciating that this would be inappropriate for their particular staff and business activity. The technology of noise reduction has improved over the years, but not long ago, the only solution might have been to enlarge the sizes of the air ducts, so that the air could flow more slowly. To provide enough space for these air ducts might require suspended ceilings, shown in the design, to be lowered and recessed light fittings to be repositioned, or replaced by projecting fittings. At worst, the floor to floor height might need to be increased leading to changes in the appearance of elevations and the design of staircases.

A similar problem has occurred in recent years with under-floor communication cabling. A number of buildings were designed without it, or with too little space allowed, leading to a later revision of floor to floor heights. On other occasions, too much space was allowed, only to be reduced at a late stage, to save money.

Another common fault has been to give insufficient consideration to the mass and size of water tanks and the space needed for pipes and access. Frequently structural calculations have had to be reworked to correct such faults.

Figure 2.2 (page 15) can be used to represent these situations. For example, if Process 3 stands for the choice of the floor to floor height of a commercial building, Processes 1 and 2 can stand for a number of inputs, including the size of air ducts, the depth of structural beams and the space needed below floors for cabling. Processes 4 and 5 then represent the dependent areas of design, for example, the appearance of external elevations and the layout of staircases. Any fault spreads through the links between design processes, so delay in discovering a fault tends to magnify the design work that may have to be redone.

In principle, the paths taken by a fault could be traced through a design work plan, to show which subsequent work might be affected by it. However, in practice, design work is not planned in sufficient detail to permit this potentially useful analysis and the knock-on effects of faults can lie hidden for a very long time.

One area that frequently gives problems on site is that insufficient tolerances may be allowed for the positioning of adjacent elements, leading to adjustments to the setting out of cladding, ceilings, tiling, balustrades etc. at the last minute. This particular problem has sometimes been made worse, where computers were used to produce the drawings and builders could not work to the precise dimensions that the computer calculated.

2.11   ORGANISATION AND PROCESS CONTROL

Just as the designers build up a model of the proposed building, these elemental processes can be considered in advance and built up into a model of how the design work is expected to proceed. This is the design programme. To devise this programme, the elemental processes have first to be defined, so that the designers know what they are expected to produce, even though the product may be no more than an abstract idea.

While designers generally work systematically, their systems can be highly individual and may even appear haphazard to casual observers. There is little doubt that effective work in a design team depends on the existence of empathy between its members, and team-balancing is as important in this area of work as any other. These aspects combine to limit the potential usefulness of planning and monitoring techniques. There is no point in planning in great detail, either where individual output might be impaired if they are not free to work in their own way, or if the co-operation between the members of a design team is so good that they will co-ordinate their work informally and actively support one another in achieving quality acceptance criteria.

Nevertheless, in design production, there has to be a clear organisation structure, to co-ordinate the contributions of different people who may never have worked together before. The design team needs to know who allocates people to tasks, who sets the acceptance criteria and who has authority to accept output, or reject it and insist on changes if these are needed. A communication structure follows from this, operating through meetings and informal contacts as well as document and data transfers.

Perhaps the most crucial aspect of the working method is the way in which design decisions are submitted for consideration or approval, to the client, other sections of a design team, statutory authorities and others. This exemplifies the general dependence of design work on good clear communication. What may be less obvious is that effective communication often depends on conveying how and why certain design options (or acceptance criteria) are favoured. Ideally, a succinct commentary about the rationale of decisions should be developed and communicated along with the design proposals.

The design programme can be of considerable help in anticipating the need for these communications, because it outlines the design processes and the sequence and time-scale in which they should take place. This framework answers questions such as: who?, how?, why? and when? Awareness of the connections between tasks also helps in defining where it is critically important to measure output against standards, as the design develops.

SUMMARY

Planning and monitoring assist the control of production processes. The inputs to processes can be adjusted to regulate their output, which is measured to ensure that it meets predetermined standards.

Design production comprises a large number of inter-linked transformation processes. Although design work is largely unrepetitive, the main principles of process control still apply. Design work aims to meet acceptance criteria, which should be set out in advance, where possible. The project brief is therefore a key document, as the acceptance criteria in all areas of a design are selected to comply with this statement. The mind-set of the design team is another key input, which may need attention.

Design work programs are structured by the essential patterns of linkage between elemental design processes. These linkages largely comprise communications, which are therefore the primary target for quality control. The project organisation will also be structured around these communications. The authority to approve output is a key element in this organisation structure.

If faults in the output of design tasks are not recognised, these can be passed on through successive steps in the work. Some consequent effects may be rooted out, but some faults are likely to remain undetected and may be incorporated in the building.

The methodology of quality management involves extensive considerations and techniques, most of which are beyond the scope of this book. However, it is significant to planning and monitoring, for example, as follows:

- Time has to be allowed in work programmes for quality control documentation to be produced and procedures to be operated.
- If the quality of work and communication is not controlled, it is likely that some will have to be repeated, and this may render the work programme inoperable.
- Information about the control of quality during construction forms a large section of design output and the processes for generating this form a significant part of the design programme for any building.
- Planning and monitoring procedures should also be the subject of quality control.

CHAPTER 3

WHY DESIGN WORK IS DIFFICULT TO PLAN AND MONITOR

3.1 INTRODUCTION

Several factors contribute to the difficulty of planning and monitoring building design work. One is a kind of 'entropy' principle, whereby it is not possible to create order in one place without creating at least as much disorder somewhere else! Designers reduce the potential for chaos in construction operations by setting out clearly what has to be done. In effect, they try to build the building in advance, using models that can be discussed between the designers and with building experts, clients, authorities, the project sponsors and others, in order to eliminate bad aspects of the design before work begins on site. In this way, the design team should take the potential for confusion during construction upon themselves.

As in the previous chapter, the term model refers to all sorts of drawings, calculations, computer realisations and similar techniques that are used to explore, test and communicate design ideas. The point of using a model, rather than starting to build straight away, is that many alternatives can be tried out and, if mistakes are made, they can be corrected much more cheaply than if a part of the building had to be ripped out and reconstructed. By this means, the process of preparing a design should remove the majority of uncertainties from the construction operations.

During design development, ideas that exist initially in people's minds are converted into models that can be understood by others. To begin with, everyone involved in the process has different ideas and many different models will be created to represent parts of the building, or aspects of the whole building. The architect will model solids and spaces, while the structural engineer will investigate the ground and suggest how to support the weight of the building, its content and the forces that may act upon it, such a wind loads. Services engineers sketch out the routes of pipes, ducts and cables, and suggest sizes and types of equipment needed. Cost consultants model the likely cost of the designers' proposals. The client may model what should happen in the completed building and how this activity will be financed and operated. Since these models are generated by people who have different interests and viewpoints, they are never perfectly compatible at the beginning. Design development is largely the process of reconciling these different models, so that the builders will be given one complete and co-ordinated model, in the form of design documentation.

The process of reconciling different partial models of the building is often made more complicated by the growth of the project organisation over a period of time. In an ideal situation, a full team of designers would be assembled at the outset, so that they can each bring their ideas and specialised knowledge together at once. In practice, the various members of the design team join it at different times and the process of developing and reconciling all the models, to make one integrated design for the building, can be long and unpredictable.

3.2   THE PRODUCTION PROCESS OF DESIGN IS ILL-DEFINED

Design is not like solving a puzzle, where there are only a few possible solutions. Instead, there may be no limit to the alternative acceptable outcomes. This can be contrasted with the builder's job, where materials can be ordered on the basis of drawings, specifications and schedules which show, at the outset, precisely what is required. Construction progress can be monitored by measuring physically complete work and materials on the site or in store. Here, most problems become visible, such as the example given in section 2.3 (see page 13), where a window frame did not fit the opening that was built for it.

Design processes lack such precise definition. It is difficult to write an extensive list of design tasks in advance and still more difficult to predefine the interactions between specialists that will be needed to co-ordinate the design solutions. When work begins, any design idea may be displaced by others that follow. Towards the end of work, it may be hard to gauge progress from the piles of drawings and other design information, since the degree of co-ordination between any two documents can be extremely time consuming to ascertain.

3.3   DESIGN PHASES

All the many aspects of a typical design problem can not be addressed at once. The work is divided between different specialist designers, and their ideas are developed and integrated through different phases and towards increasing levels of detail.

Figure 3.1 illustrates how there is a gradual shift of emphasis as design work progresses through these phases. The initial task is to analyse the purposes of the project and the major elements of the design. The acceptance criteria then have to be clarified, so that choices can be made between alternative design proposals. Typically, this is done at the same time as initial design work, because the criteria may be understood more clearly when related to actual design proposals. Following this, the best design options are developed in detail and adopted as the solution.

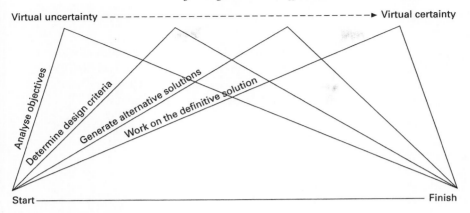

Virtual uncertainty – → Virtual certainty

Analyse objectives

Determine design criteria

Generate alternative solutions

Work on the definitive solution

Start ——————————————————————— Finish

Fig. 3.1 The changing emphasis of design work
Source: Diagram reproduced from the PhD dissertation 'Bouwkundig Ontwerpen – een beschrijving van de structuur van bouwkundige ontwerpprocessen' submitted to Eindhoven Technical University by Jan Thijs Boekholt, in the 1980s.

3.4 DESIGN STAGES

Various professional institutions have published a formalised view of the main stages of design work, in an attempt to make it more controllable. The key difference between the concepts of phases and stages is that the latter should be amenable to fairly sharp definition, as to when one stage ends and the next begins, whereas the former accepts that design work cannot be divided so neatly.

By defining the stages of design work, the institutions also seek to clarify what tasks should be done, as the design develops, and the output that their members are expected to produce. This not only helps the design practices to list the 'deliverables' that must be output at each stage, but also helps them to claim fees in stages, for identifiable products.

The professional institutions of the engineering disciplines, landscape architects, etc., each divide design development in different ways, but all their approaches distinguish between:

1. *Project Definition*, which establishes the aims of design work and generally precedes it.
2. *The generation of outline design proposals*, for analysis and discussion.
3. *Scheme design*, which provides sufficient preliminary information for the client and the planning authority to authorise the building project to proceed.
4. *Detail design development*, which ensures that the designs of the various engineers and specialist consultants (and sometimes suppliers) are fully co-ordinated. Information produced at this stage is generally sufficient to obtain approval to proceed to construction, under the building regulations and other statutory instruments.

5. *Production information*, which is the most detailed level of design and describes the 'nuts and bolts' of the construction.

One of the most widely known definitions of work stages is the plan of work, which was originally published by the Royal Institute of British Architects (RIBA) in 1965.[1] This divides the entire building project into stages. These may be broadly grouped under the headings used by the other institutions, as follows:

Project Definition
A   Inception
B   Feasibility
Outline Design
C   Outline proposals
Scheme Design
D   Scheme design
Detail Design
E   Detail design
Production Information
F   Production information
G   Bills of quantities

Further stages then relate to procurement activities (appointing contractors), the administration of construction work and, finally, the hand-over and occupation of the completed building:

H   Tender action
J   Project planning
K   Operations on site
L   Completion
M   Feedback

Figure 3.1 (page 25) helps to illustrate that the work cannot be as neatly wrapped up into the stages that were suggested by the RIBA and the other institutions. The phases it refers to may occur within any of the design stages, or they may continue across several stages, depending on which part of the building is being designed and when specialised designers are brought in to the design process.

For example, consider the RIBA work stage definitions. In many projects, the majority of the production information is generated by specialist contractors, after construction contracts have been tendered. This means that part of work stage F actually takes place simultaneously with stage K. Another case in point: recent legislation, pertaining to safety on site, has highlighted that project planning has to commence much earlier in the design process, and proceed interactively with it. Thus, stage J should be taking place whilst stages D, E and F are in progress. Similarly, cost control must be carried out interactively with design development and the bill of quantities, referred to as stage G, is just one (optional) deliverable in this process.

If the stage, or phase, definitions used by different institutions are looked at in detail, it may be found that their boundary definitions do not precisely correspond.

3.5   ABSTRACT DESIGN ELEMENTS

Many design problems are not related to the physical elements of a building, but to potential interactions between them and also to the people who will construct or use the building.

For example, the design team should take account of:

- Movements of people, materials, work and activities to be accommodated.
- The 'readability' of spaces, including how easy it is to find one's way around.
- Safety aspects, during construction and when the building is in use, both for the occupiers and people who may be engaged on cleaning and maintenance work.
- The amount and quality of lighting.
- Aesthetic continuity, including colour and texture.
- Noise generation and transmission.
- Structural stability, including potential failure modes and the characteristics of assemblies when partly built.
- Air movement, as this affects heating, ventilation and the possible (and unpredictable) effects of a fire.
- The possible effects of poor workmanship, weathering and vandalism.
- Maintenance requirements and running costs.

3.6   DESIGN CYCLES

Designs generally develop in a cyclical way, as shown in Figure 3.2. Due to the complexity of building design problems, particular specialist designers focus on different parts or aspects of them. For example, engineers concentrate on the structure or building services, while the architect looks at the use of space, lighting, fixtures, furnishings and the surface finishes. While requirements are being established for each part or aspect of the building, such as the elevations, patterns of movement within the building, heating and ventilating systems, special areas, such as kitchens and toilets, each design discipline may work in relative isolation. The partial or preliminary designs that each proposes are then compared and modified to create an integrated design for the building.

To interpret Figure 3.2 correctly, one should imagine that several such 'loops' are superimposed. The requirements and options for several elements and aspects of the building are first explored. Then, at the bottom right corner, the partial solutions (or preliminary designs) are compared and an attempt is made to integrate them into a design solution that works in every respect. Usually, there are mismatches and some, if not all, of the work has to be reconsidered and adjusted. This process of revision and refinement is referred to as 'iteration' and it is a normal method of working. Since it is known in advance that iteration will occur, design work planning should take this into account. The difficulty is that no-one knows in advance exactly how many times a particular elemental design will need to be iterated.

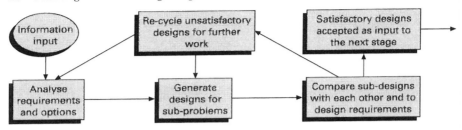

Fig. 3.2   Typical loop in the progress of design work (iteration)

The acceptability of partial designs can be very difficult to appreciate before all comparisons have been made with other partial designs. Such comparison is only possible as the design output is nearing completion. Typically, inconsistencies between aspects of preliminary designs come to light little by little, especially when the output of different design specialisms is compared. One particular skill required to manage a design process efficiently is, therefore, the ability to recognise areas of a design that can be developed independently, without a high risk that they will have to be reworked after they have been compared with other parts of the developing design.

For every aspect and area of the design, there may be thousands of tasks to do to reach an integrated solution. The detailed flow of design ideas and communication between the designers cannot be predicted, nor can the number, frequency and subject matter of design cycles (or iterations) be predefined. There are, however, recognisable patterns that can be used to plan design work and monitor its progress. These are discussed in the following chapters.

3.7 BACK-TRACKING

Design development gradually adds more and more detail to previous decisions. There may be occasions when design decisions, made at an early stage, have to be revised to accommodate the advice of specialists who are drafted into the project team later on. Where the need to rework designs arises unexpectedly, this is referred to as back-tracking.

The client, the financiers of the project, authorities (including conservation groups) and the various design disciplines will each have their own view of priorities for a particular construction project. Although every effort may be made to analyse and discuss these during the briefing, differences of opinion or interpretation are likely to arise as the actual design emerges. Many designs are done in the belief that they will be acceptable, whereas another member of the project team may later insist that further thought is given to them. Designers are often asked to reconsider the options, in order to come up with solutions that satisfy a higher standard, or a wider range of criteria.

Back-tracking is distinct from iteration in that it requires unplanned changes to be made to work that was previously thought to be good. It may be helpful to look

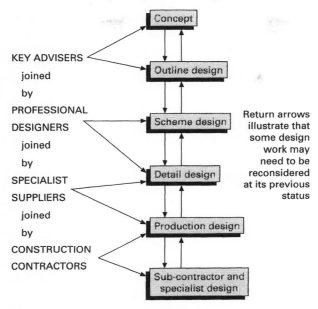

Fig. 3.3 Back-tracking as the design team grows

again at Figure 3.1 (page 25). Designs prepared at an early stage may be produced on the basis of design criteria that have not been worked out in detail, or sufficient precision. As work proceeds, these uncertainties are gradually ironed out so that work proceeds on a firmer footing. It is convenient to think in terms of status. The status of design work changes, as preliminary designs either become accepted as working solutions, or are replaced by more acceptable designs. Work that is produced late in design development tends to be more conclusive and eventually achieves the status of 'approved for construction'. Considering that the profiles shown in Figure 3.1 can apply separately to different aspects of the design, and the successive phases of design development, the overall situation is often quite complex, whereby designs may have final status in one context, but be only preliminary in another. The result is that the status of design output may quite often be set back, in the view of one discipline or another, from a complete to an unfinished state.

Figure 3.3 illustrates the tendency for back-tracking to occur as new specialists are added to the design team. It can also happen, of course, if mistakes occur in the work of one design specialism, which are not recognised before other work is done, using this as input information.

It becomes increasingly difficult to unravel design decisions, as they become enmeshed into the growing set of co-ordinated drawings and data. Back-tracking on previous decisions tends to become more difficult and time-consuming as design work nears completion. It is therefore something to be avoided.

This is where the greatest difference between the design phase and the constructional operation lies. In design work, what appears to be complete and finished (or at least, reasonably well advanced) may, subsequently, prove not to be so, whereas in construction work instructions are generally known in advance and progress can be measured by the physical existence of the relevant parts of the building.

3.8   DIFFICULTIES IN MEASURING OUTPUT

Because design work progresses through overlapping phases, as shown in Figure 3.1 (page 25), different elements of the design can arrive at stage completions at different times. Furthermore, the complex relationships between different building elements that can generate and regenerate iterations of design tasks, as shown in Figures 3.2 and 3.3, make it difficult to be sure that any part of the design is finished until the entire design is complete.

Figure 3.4 contrasts the input and output phases of a design stage. During the latter part of the work, progress can be measured as the sum of complete design output, that is, drawings, schedules and specifications. However, the quality of co-ordination, between a very large number of drawings and specifications, may be very difficult to ascertain.

Output during the first half of the design programme is less in volume and tends to to be sketchy. A true picture of progress can therefore be difficult to establish. Nonetheless, progress in the early part of a design stage is very important, as this time has to be well used in co-ordinating the requirements of the different design specialists. To make certain that later work will not be hampered by an uncertain

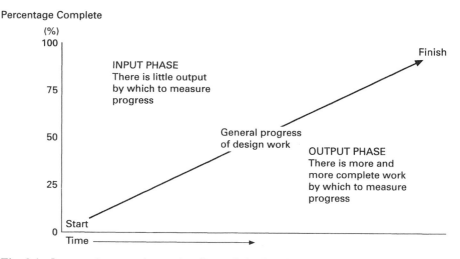

Fig. 3.4   Input and output phases of outline and detail design work

start to the work, it may be necessary to analyse this phase of the work in detail and define outputs that can be recognised and tested against acceptance criteria, to be sure that real progress has been made.

3.9   FRAGMENTATION

Design work is fragmented in many ways. Designers work in different offices and they are drafted into projects at different stages. Consultants and specialist sub-contractors have different contractual relationships with the client and construction companies. It follows that the interest in design work planning and monitoring varies.

Design models are also fragmented, as information is held in different locations and forms. Specifications, schedules, conceptual and detail design drawings, the cost plan, and construction working methods all model aspects of the proposed building operation. Each is different in its structure and likely to have been produced by different people, even if they come from the same office. It can be extremely difficult to ensure that the overall objectives of the project are effectively translated in each of these media, and that the story told by each one is co-ordinated with and supported by the others.

The media used to carry the design output, for example pieces of paper and computer files, offer no mechanical means of ensuring that they are co-ordinated.

Organisations that stand outside the design process, such as the statutory authorities, can also have unpredictable influences on design decisions, for example, by rejecting outline designs, or by limiting the site access or the availability of services.

3.10   INNOVATION

Innovation is far more common in building design than people usually appreciate. Most buildings incorporate hundreds of different components and, in all but the most repetitive building types, such as mass housing, the combination of components is invariably unique. The configuration of the building is also likely to be unique, on a unique site. There will always be untried connections and unrecognised relationships between these factors and between components. This renders it impossible to ensure that every aspect of design work will be right the first time it is done. A great deal of imagination is needed to evaluate designs as they come off the drawing board. To get the inputs to innovative design right first time requires considerable intelligence and careful framing of the design aims and parameters, based on circumspect observation and careful research. The fact is that this can never be done well enough. Since mistakes will, inevitably, be made, designers have to rely to some extent on their instincts and judgements. This is, after all, an essential aspect of expert capability.

3.11 DECISION-MAKING STYLES

Construction companies (and many clients) have well formulated tasks to perform that can be planned in detail, but designers often have to research the problems they have been set, before these become open to solution.

Strategies

Various strategies are adopted by different designers in response to ill-structured problems. Three typical approaches are described below.

Attempt to structure the problem as rationally as possible

Designers who work in this way concentrate their initial efforts upon getting the maximum information about their client's requirements and other similar buildings that already exist, before committing themselves to a design. These designers can be difficult to work with because feedback from them is at a premium during the early stages in that they find it hard to present preliminary ideas for discussion.

Inspired foresight

This can be a very successful strategy in the hands of experienced designers, who get the feel of a problem at the outset, from a very slim brief. These 'impresarios' develop a powerful concept straight away and are then able to bring in the details of the design without compromising their original idea. A problem that may be encountered with these designers is that they appear self-opinionated and hard to influence.

Structure the process of developing a solution by following a series of steps

With this approach the initial design work may proceed by:

1. Developing the brief by questionnaires about the building functions.
2. Using techniques, such as relationship matrices, to analyse functional and spacial relationships.
3. Analysing and grouping building services requirements.
4. Analysing the functions of partitions to determine the types required and their locations.
5. Defining the structural requirements and exploring options that satisfy them.
6. Defining and refining the building envelope.

The work of designers who structure the process may be more amenable to planning and monitoring than those who work intuitively. Typically, structural and mechanical engineers work in a structured way, with calculation as a principal design tool, whereas architects rely more on drawing techniques. It has been suggested that structured processes of design inhibit creativity and that architecture created this way may lack the inspirational qualities that can typify the first two approaches listed above.

The appropriate choice of designers may, therefore, depend not only on the priorities of time, cost and quality, but also on the *kind of quality* that is wanted, whether a workmanlike solution is required or an inspired building design is sought. The ways of working with which the client and project manager feel comfortable, may also dictate the choice of designer. Staff in a multi-disciplinary team may have to accept that they will need to form different working relationships to suit particular designers and to accept output in various forms (drawn or calculated, sketchy or engineered) to keep themselves informed about the progress of designs and communicate effectively.

3.12   A MISSING BODY OF KNOWLEDGE

The implementation of systematic planning and monitoring can be hampered by a 'Catch 22' situation.[2] If the managers and designers lack experience of control by planning and monitoring it is likely that work will not proceed according to plan and the inevitable failure will damage the credibility, not only of planning and monitoring systems but also of the individuals who attempt this approach. The effect of this may be to inhibit the development of planning and monitoring techniques.

As a consequence of fees for design work being reduced, in response to competition, design practices may find that they lack the resources to plan and monitor work in detail. Partners (or practice managers) are understandably reluctant to increase their overheads with activities, such as this, if they are not obviously productive. They may even lack the time to find out about the potential benefits and methodology of planning and monitoring.

If the plan for any stage of design work lacks detail, progress monitoring may be ineffective. Without monitoring, there is no way to anticipate any problems that may arise and adjust the plan to avoid or minimise them. If the detail is insufficient, it is difficult to explain to the design team and the client, alike, why certain design tasks should be done and decisions taken at that particular time, rather than later. There is also an associated need to recognise tasks and decisions that should be held over until the necessary information becomes available. Even decisions of critical importance, such as appointing all the necessary design consultants, can be seriously mistimed if the durations and precedence of design tasks are not made clear early enough.

SUMMARY

One function of design work is to enable construction to proceed in a systematic, and therefore economical, way. If design work is not planned, this process is liable to take on the confusion that would otherwise occur during construction, by generating unpredictable patterns of iteration (revision and refinement) and back-tracking on earlier decisions.

During design development, there is a gradual shift of emphasis from considering objectives and acceptance criteria, towards the generation, evaluation and integration of design solutions.

Acceptance criteria are rarely cut and dried at the outset of work, but tend to develop and change as work proceeds. This also adds to the difficulty of measuring the quality and completeness of design output.

Designers in the various specialised areas generate preliminary solutions, which they model in various ways, including drawings, calculations and specifications. These studies are then reconciled with those produced by other designers, into one integrated design.

The quality of co-ordination between elements of the design may be difficult to establish at any given time. The problem of comparing designs is further complicated where the various design models are represented in different forms and media.

Design work, on a typical project, is done by several people who are working for different companies and in different locations and, for this reason, their activities can be difficult to monitor and co-ordinate. Efficient production planning requires an appreciation of how far the different areas of a design can be developed independently of one another.

Every design incorporates combinations and configurations of materials, components and assemblies that have not been put together in that way before. The time and skill needed to control the quality of this innovative work can be unpredictable.

Whereas the output of work provides many physical products by which to measure progress, monitoring points may be lacking during the input phase of each design stage, so that progress can not easily be gauged.

Designers' decision-making styles may differ, from each other and from that of the client, financiers and others involved with a project.

The common lack of experience of planning and monitoring methods can make it difficult to plan work and monitor progress in an effective way.

NOTES AND REFERENCES

1. Royal Institute of British Architects (1965, reprinted 1983) *Plan of work for Design Team Operation.*
2. *Catch 22* is a novel by Joseph Heller, in which American pilots, stationed in the Mediterranean during the Second World War, found themselves unable to escape combat missions except through serious injury.

CHAPTER 4

BASICS OF PLANNING AND MONITORING

4.1 INTRODUCTION

Before it is possible to describe the detailed methodology of planning and monitoring design work, it is necessary to consider some general principles and techniques, which could be applied to any type of project.

Broadly, work planning proceeds by four stages:

1. Work breakdown – identifying what needs to be done.
2. Estimating the duration of each identified task.
3. Optimising the sequence of key tasks to calculate the minimum duration of the entire work programme.
4. Reconciling this analysis with the actual availability of time and staff and other resources.

This establishes what is referred to as a 'base plan'.

Network analysis is an essential tool in optimising the sequence of operations and its application to design work is outlined below. However, no advice is given here about how activity networks are calculated, because this is clearly explained in many other books.[1]

To keep track of work as it progresses, it is necessary to:

- Measure what has been done.
- Compare actual progress with the base plan.
- Consider how much (calendar) time and resources are still available to complete the work yet to be done.
- If progress drifts significantly away from the base plan, the plan of work may have to be revised and the new version agreed with the design team.

The term 'resources' can be used to refer to available staff, equipment and materials. It is often convenient to state the value of all these resources, as a group. In the case of design work, available resources can be equated with the unspent portion of anticipated fees.

4.2   PLANNING FROM THE TOP DOWN

It is not usual to plan an entire project in detail at the outset, as this would require considerable effort and show very little immediate return. In the case of design work, details that should be taken into account in the plan may only become known when the various consultants and sub-contractors are brought into the project team. All that is really needed in the early stages of the project is a 'master plan', to show when the most important stages of the work should begin and when they should be completed. As each stage starts, the work of that stage is considered in detail, usually with help from managers of the specialist design companies or departments that will be involved.

Work planning proceeds in a hierarchical structure, whereby the large groups of operations – the 'top' of the planning system – are identified first and these are later analysed downwards in terms of component tasks and the detailed relationships of activities.

Monitoring, as explained in section 4.15 (page 51), works in the opposite direction, from the bottom up.

4.3   BASE PLANS

Progress can only be measured against a work plan. There will be several different levels of planning and each must be approved by a different level of the management hierarchy, as appropriate.

At the top of the planning logic is a project master plan, which would be agreed at the highest level. Figure 4.1 shows a master plan for a building project, in the form of a simple bar chart. In this figure months are numbered, whereas most project master plans would show actual calendar dates. Since design work precedes construction, the project master plan is normally agreed between the design practice partners and their client, even where a project manager is also appointed.

More detailed plans would be developed for each stage of the work shown in the master plan. Those required during the design work would be developed between the leading designers in each professional discipline and agreed with practice partners. These detailed plans provide the basis for co-ordinating related areas of work and monitoring its progress. For example, there should be a separate plan for the whole of the outline design work and another for the detail design work. It is not unusual for these two design stages to be separated by lengthy planning approvals procedures, or while the client completes the financial arrangements for the project. During this delay, the members of the design team may change, or the client may decide to appoint a completely different design team for the detail work. Thus, the work planning for detailed design work may be carried out by different people from those involved at the outline design stage. The base plan then forms the basis of agreements about who should contribute what work to the project and when. Such plans can thereby form the basis of co-ordinating the work done by different people and companies.

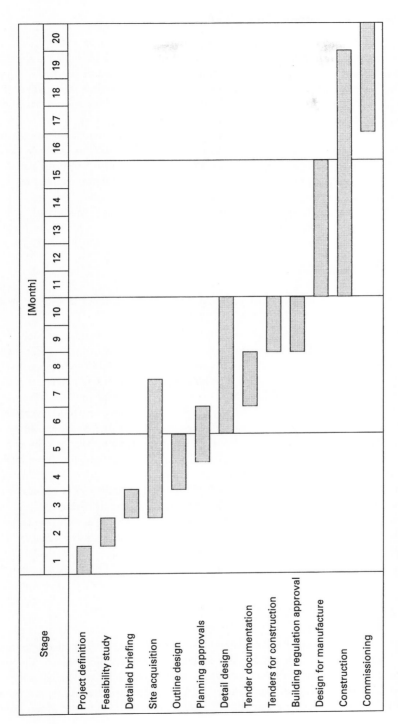

Fig. 4.1 A project master plan (illustrative only)

There is a second reason why such plans are referred to as base plans. In practice, work programmes are often revised and reissued; the base plan is then needed as a fixed datum against which actual progress can be measured.

At the completion of the work, it is prudent to archive the base plans, along with the records of actual progress, for future reference and as a guide to improving the accuracy and credibility of work planning on future jobs.

When construction contracts are let in the traditional way, tendering procedures separate the design and construction phases. In contrast, the start of some work packages on a fast-track project will overlap the completion of other areas of the design. In this instance, design work has to be shown on the same plan as the procurement programme and the construction operations, because these are interrelated. A typical design process programme for a fast-track project is illustrated in the Appendix (page 180).

4.4   WORK BREAKDOWN STRUCTURE

Work may be considered in several different ways, to ensure that planning takes all relevant factors into account, the main three being:

1. The major stages of the project, and the tasks that need to be done in each.
2. The people involved in the work and the relationships between them. Figure 4.2 shows an organisation chart of the structure for a building design project. In design work, the main specialisms are architecture, structural engineering and services engineering. In addition, a wide range of other skills may be called upon, from urban planners to kitchen designers. It is important to appreciate that many significant tasks are done by people other than the designers. For example, clients contribute to the briefing, approvals have to be issued by the statutory authorities and the public utilities may supply designs for the main services leading to and from a building.
3. The products that have to be created. In design work, this category covers the drawings, specifications, schedules, calculations and project-related documentation, such as the brief and reports that are needed at various stages of the work, to describe the rationale of the project and explain the basis of recommendations and what is being designed.

Broadly, these 'dimensions' of the work breakdown structure correspond to:

1. The design production process.
2. The design organisation, including the communication network.
3. The deliverables, i.e. the products of work.

The production process and the deliverables can conveniently be planned together against a time-scale, since most processes result in an identifiable product. The organisation of resources must be considered together with this analysis, as this largely determines the rate of production and, hence, the duration of operations. In the case of design work, the availability of designers is the main factor which limits how quickly work can be done.

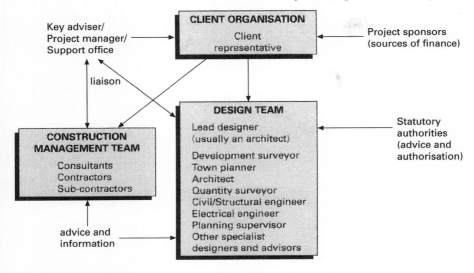

Fig. 4.2   Typical organisation structure for a building design project

4.5   SEQUENCE OF OPERATIONS – TASK DEPENDENCY NETWORKS

The work of each substantial stage of design and construction operations is carried out in a number of steps that are referred to, in planning terminology, as tasks or activities. It is important that these activities should be carried out in a sequence that will achieve both efficiency and the required quality of work.

As each stage begins, the necessary tasks are listed and the best sequence in which to carry them out is considered. In practice, the overall approach to design work usually follows a pattern that is well understood by the members of the design team, from their training and previous experience. Five common approaches to the the procurement of construction services are discussed in Chapter 6, in relation to the influence this may have on the planning of design work. In each case, the detailed sequences of design work, and the communications that connect them, need careful forethought if unexpected delays are to avoided. In practice, these methods are frequently amended to suit the requirements of a particular project.

Once the necessary tasks (or activities) in a work stage have been listed, the next step in formulating a rational work plan is to arrange these in the order that they must be done. Very often, it is found that the product of one task does not simply flow as the only input to the next, but that the products of tasks are utilised as input to several others. Similarly, tasks often require the product of several others, as input. The result is that elemental processes are not simply linked as a chain, but form a network of interdependent activities.

Many tools and techniques are available to help the work planner. The following introduces some of those which are most commonly used. The examples are chosen to gradually build up a picture of what design work programmes can look like.

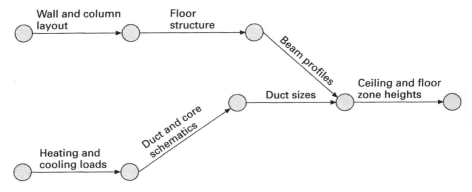

Fig. 4.3   Part of a planning network, with activities shown on the arrows

4.5.1   Precedence diagram – with activities shown 'on the arrow'

Figure 4.3 represents identified tasks as arrows. Each arrow comes from the previous task, or tasks, on which it depends and goes towards the tasks that can only be done when it is finished. The activities in this diagram represent a fragment of the design process for an office building. To see how it works, consider the reference to ceiling and floor zone heights, at the right of the diagram. The overall dimension of the ceiling void, structural and false floor would be decided by the architect. It would depend on the structural engineer's advice about the depth of beam profiles, the above-ceiling duct sizes, as calculated by the services engineer and the space to be allowed for communication cables, between the structural floor and the actual office floor. In this example, the false floor is ignored.

The two information inputs are shown as arrows leading into the architect's decision about the floor zone height. This diagram follows the general convention that earlier work is shown to the left of work that follows later. The progression of work flow can also be shown starting at the top and going down the sheet, to end at the bottom. It is possible to distinguish between the activities of different design specialisms by the use of different colours, for the arrows or the text.

The circles at the ends of the arrows, in Figure 4.3, represent what are called events. Events in design work may generally be recognised by the transfer of information, as needed for the work that follows. While arrows suggest that activities are continuous, in practice, events can actually be characterised by delays, for example, until a member of the design team is free to work on the following task.

Networks with activities shown on the arrow can help to sort out the detail of relationships between limited numbers of tasks, but where many activities must be shown, this particular method for representing the work flow has some complications. For example, 'dummy' activities are sometimes needed to connect activities, because the arrows can only start and end in one place.

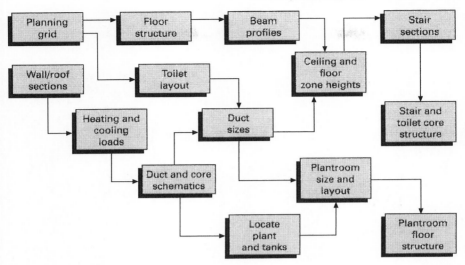

Fig. 4.4 A network of interdependent design tasks

4.5.2 Precedence diagram – with activities shown on the node

Figure 4.4 shows a network that covers an extensive area of a design process, at a relatively high level (i.e. without much detail). In this example, the activities are shown as boxes, which form the nodes, or connecting points, of the activity network. This form of representation is free of dummy activities and each arrow represents a particular transfer of design information. Time is not truly indicated and the boxes may be placed in a different sequence from the work, provided that the arrows always go from the task that has to be done first to the one that must be done later.

This representation only shows the precedence of activities, that is, which tasks follow which. In practice, a lot of useful information can be added within, or adjacent to the boxes. This information typically includes: how long each task is expected to take, the earliest and the latest time it can start and the earliest and latest times when it is expected to finish. The tasks of the different design disciplines can be differentiated by the use of colours, or by giving the boxes of each discipline a different shape.

4.5.3 Precedence diagram – Gantt chart

Figure 4.5 extends the small fragment of the design activity network shown in Figure 4.3 (page 40), this time in the form of a linked bar chart. This representation is often referred to by the name of its inventor, Henry Laurence Gantt. Gantt lived in the US and worked for Frederic Winslow Taylor, a well known pioneer of management science. The linked bar chart shows the planned

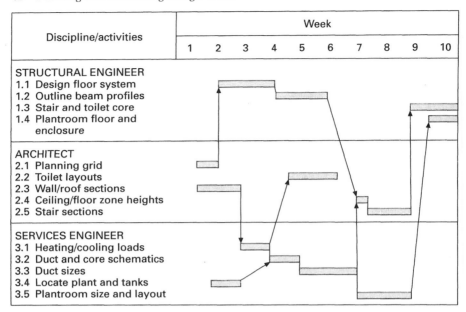

Fig. 4.5   A partial design programme (from a detail design stage)

duration of each task by representing these as bars, with a time-scale of weeks shown at the top of the chart. This is a considerable advantage over showing activities on arrows or nodes, since the chart instantly conveys which tasks follow which and which may proceed at the same time.

In Figure 4.5, the activities of each discipline are grouped together. This makes it easy to see where information should pass from one design discipline to another (without the use of colour). In practice, the dependencies between tasks within each discipline far outnumber the links between the different design specialisms. Information dependencies internal to each design discipline can generally be managed by the particular specialist team, without external planning and monitoring. The information dependencies between the disciplines are generally more significant to overall progress and it is helpful to highlight them in some way. For example, if one design office is falling behind in their work, a chart in this form makes it relatively easy to pick out which activities in the other design offices are likely to be held up as a consequence.

4.6   SELECTING ACTIVITIES TO INCLUDE IN THE ANALYSIS

It would be possible to analyse design work in great detail, but this is not done, for several reasons. Since the patterns of necessary iteration are unpredictable, designers can not, in fact, work to a plan that shows a lot of detail. There is also a limit to the value of detail, as it is only worthwhile to plan to a level where the design team actually learns something from the analysis. The most useful knowledge to come out of planning is likely to be an appreciation of how each designer's work is dependent on the information and ideas produced by others, especially those working in the other disciplines.

Since less than one-fifth of the connections between design tasks pass between disciplines, only about 20 per cent of the total interdependency of design work needs to be considered in detail, when planning and monitoring the work. However, if too few activities are shown in the planning network, the longest sequence of interdependent tasks may be underestimated. This could make it hard to predict when difficulties could arise, in co-ordinating activity.

The appropriate selection and definition of tasks, for the purposes of planning and progress monitoring, is likely to improve with experience. It is advisable for the person, or group, responsible for planning a design programme, to discuss the selection of these tasks with the leading designers in each discipline, thereby drawing on their experience. The key questions to ask may be: 'On what information, from the other disciplines, does your work most depend?', 'What has held your work up in previous jobs?', and 'What is different about this particular job; what could catch us out?' Knowledge of these significant linkages is important in developing the capability to group design activities, which may actually have very many interdependencies. The network of dependencies is kept as simple as possible to understand, so that when it is analysed, discussed with the team and agreed as a programme, it is effective as the key reference for ensuring that work proceeds in a smooth, co-ordinated and well motivated way.

4.7 CRITICAL ACTIVITIES AND CRITICAL PATHS

On every project there are certain tasks which must be done promptly so that chains of subsequent activity are not held up, whereas delay to some other tasks may have little effect on the programme. Advance planning should identify these critical tasks. Monitoring should then ensure that these are carried out as soon as the right information and people become available to do them.

The minimum duration of each stage of work depends on a sequence of critical tasks, which is known as the 'critical path'. Each task in this sequence depends on output from the preceding task, as input. Whereas every task in a design activity network, except those at the very beginning, is dependent on information from earlier work, the critical path is special, because, when the expected duration of every task in this sequence is added together, this total is greater than any other sequence in the network. Thus, the critical path is the sequence of inter-dependent activities that is likely to take the most time.

Although Figure 4.5 (page 42) does not show iterations, it does show a critical path. This can be picked out by the presence of vertical arrows. These connect activities that follow one another without pause. Where the arrows link activities at an angle, this shows that there is a margin of time between them.

When the critical path through a work stage is first identified, and its total duration is calculated, more often than not this total exceeds the time available for the work stage. This is particularly true of design work, where the activity network may be extended by iterations. This also makes the critical path much more difficult to recognise, because iterations are likely to occur in task sequences that have not been identified as critical, but which turn out to be critical when a task (or tasks) cannot be finished satisfactorily at the first attempt. This uncertainty, inherent in the progress of design work, means that there are likely to be several paths that should be treated as if they are critical, because one or more of them may turn out be critical, although this was not apparent in advance.

In building design work, the critical path invariably crosses and recrosses between different design specialisms. The work of the various professions in the team can not, therefore, be planned in isolation. If any of them falls behind on a critical task, the inevitable result is that the work flow in other offices will be disrupted.

Figure 4.5 also illustrates that the incidence of critical tasks within each discipline is likely to have a staccato pattern. In this example, which is only a small fragment of a complete design activity plan, the architect should work hard in weeks 1, 2, 7 and 8, in order to progress activities 2.3, 2.4 and 2.5, which are on the critical path. The services engineer should be working hard in weeks 3, 4, 5 and 6, to ensure that activities 3.1, 3.2 and 3.3 do not fall behind schedule.

Thus, intense effort may be called for, to complete the 20 per cent of tasks that are on the critical path in the time allocated to them. Members of the design team may be permitted to breathe a little more easily when they are engaged in the remaining 80 per cent of their work. It should be noted, however, that for each discipline there may be more than one activity in progress, most of the time, that could turn out to be critical, if the predicted pattern of iteration becomes extended. Key designers may therefore often find themselves under pressure, from the work programme and those who monitor it.

4.8   FLOAT

Float is the spare calendar time available to complete a task, as shown on the plan. (There is rarely much 'slack' in the available resources.) As explained above, this can be recognised in Figure 4.5 by the diagonal arrows. The activities which float are those which are not on a critical path.

Float should be treated with care. If a task is delayed until the end of the period in which it is scheduled to be done, it is liable to change into a critical task, that is, one which could delay other critical work if it is not completed on time. In design development, almost any output may require amendment, after it has been seen by other designers, the client or statutory authorities. Also, the amendments to other parts of the design can have knock-on effects leading to hasty work on tasks that were originally not considered to be critical.

So, float is not a luxury, to be frittered away. It gives essential margins in the programme. Without float, many design programmes would be unworkable.

4.9 MILESTONES

Target completion dates for major sections of the work – and, sometimes, important decisions – are often referred to as milestones. These are not spaced out equally, as the term implies, but fixed in relation to the time that the senior managers believe to be necessary to complete sections of work.

For example, in construction, completion of the weather-tight enclosure of the building can be regarded as a milestone, as this enables work to commence on the internal services and finishes that, otherwise, would be damaged by the wind and the rain.

Typical milestones in a building design programme mark the completion of key activities. These include the stages shown in the project master plan (see Figure 4.1 on page 37), but could also include events such as quality control points. This suggests a list, as follows:

- The client agrees the detailed brief (setting key design objectives).
- Site acquired (making it reasonable to commit money to design fees).
- Outline design ready for the client's approval (usually prior to applying for detailed planning approval).
- Town planning permission applied for. (This signals the completion of a significant section of the work and usually means that aspects of the building design are frozen and should not be changed.)
- Preliminary design review completed (when a substantial amount of detail design work is available for checking).
- Planning grid finalised. (This provides the key dimensional basis for integrated architectural and engineering design work, near the beginning of the detail design phase. The base grid is normally communicated by way of a set of 'base drawings' – not to be confused with the base plan, which is a work programme.)
- Building control application made. (This milestone requires sufficient detail in the design for the authorities to assess whether all the statutory requirements will be met in the finished building.)
- Critical design review. (When the integrated designs are put forward for a final check, prior to their release as construction information. It is preferred that this milestone is reached before the documentation for tenders is issued, since any necessary alterations after this time would delay construction work, bring contractual complications and incur additional expenditure for the client.)
- Release of detailed design information for construction contracts to be tendered. If the project is being managed along traditional lines, bills of quantity are prepared from this information and form an important part of tender documentation.

The release of detailed design information should be preceded by the critical design review, to ensure that a minimum of changes are likely to be made after contracts have been signed with the builder, sub-contractors and suppliers.

In fast-track programmes, the final design information is released in a series of packages. It may follow that the critical design review also has to be split into several partial checks, as each package is completed.

The criticality, or the float, of each design activity should be calculated in relation to the next milestone. Float cannot be borrowed from the next stage of work. In fast-track projects, the milestone representing the release of final information for construction may be split into several smaller ones, to correspond with the dates when work packages are released for tender.

4.10   HAMMOCKS

Hammocks are a feature of network analysis computer programs. This enables the planner to quickly move between the high level view of the project master plan and the lower level views of activity networks which detail the activities at each stage of the project. For example, where the network is presented as a Gantt chart, by selecting a bar of the master plan, a click of the mouse would bring the detailed network that underlies this bar to the screen. The term hammock is therefore an analogy, which suggests that the detailed network lies in the bar that represents it on the higher level (summary) network. Most proprietary network programs enable networks to be shown at several levels of detail; in effect, there can be hammocks within hammocks.

This enables sub-programs for sections of work to be analysed more-or-less independently. The bars of the project master plan, as in the example shown in Figure 4.1 (page 37) and also that in the Appendix (page 180), each summarise substantial areas of work that require co-ordination between disciplines, whereas the work plans in a hammock generally show tasks that are specific to one or other discipline. Figure 5.1 (page 60) exemplifies such a second level planning chart, which may be compared with the one in the Appendix (page 180). Each task at the second level may be analysed in greater detail at a yet lower level, where only the tasks of one discipline are represented. This third level constitutes the work plan for the particular discipline, during a specific stage of the project.

Depending on the particular software in use, the calculation of a work planning network can be restricted to the area being analysed, or extended to recalculate the entire project network, as each task is added in or if an alteration is made. This is important during the development of a work programme and during the monitoring of progress, because the project master plan should not be influenced by small variations that may be taking place in the detailed planning or progress of the work of any one stage or discipline. One aim of analysing the production process, or amending it, is to ensure that all the tasks of a major stage are completed before the allotted deadline (milestone). It is therefore helpful to the planner to be able to focus on selected sections of the work plan and make adjustments in that area, until the sequence of tasks and the resources allocated to them predict completion within this time frame. Clearly, this analysis is merely theoretical, unless and until the various design teams and the managers authorised to allocate staff agree to work to the plan.

Where long-lead items set design deadlines (see Chapter 1, section 1.9, page 6), these are often derived from the plan for the construction operations. This means that, in some instances, critical paths jump milestones, to connect activities in one work stage with those in another. Most network planning software deals easily with this, but the planner may need to be vigilant at times, in case work in a localised area could be having effects in other parts of the work programme. The most important thing is that the planner should be using the tool competently and have obtained realistic data from other members of the project team, on which to base the calculations. There can be a problem in relation to the construction work programme, since its details are rarely available at the time when design work is being planned. Long-lead items therefore tend to be instances where strategic thinking is needed. In turn, this is likely to influence decisions in the design itself.

If a stage or a task finishes late, or early, the software can instantly recalculate the dates when subsequent activities should take place. Such recalculation used to be quite laborious to do by hand and was, therefore, only done if essential. Modern computing software has made planning and monitoring capable of being much more flexible and responsive, as design development work moves forward. Planning has become much more interactive with the actual progress of work and has, therefore, become much more powerful in its capability to guide work towards effectiveness and efficiency.

4.11   CRITICAL LINKS AND HIGH-RISK NODES

High-risk nodes on a planning network are activities or events which could easily be delayed and would, in their turn, delay many of the following activities if they are not completed at the planned time.

Town planning applications are a good example, where designs have to be submitted in time for review by a committee that meets only periodically. The design has to be reasonably complete and well thought out before it is submitted for approval, or difficult problems may have to be resolved later, when the design is developed in more detail. Input to the design for planning approval may be required from a number of specialists and, if any one of these submits work late, or errors are discovered, the submission could be delayed. If the design, or its presentation, is not good enough, the design might be rejected by the committee and the entire project would then be delayed, by several weeks at least.

In Figure 4.4 (page 41), the activity 'duct sizes' could be a high-risk node, since neither the ceiling nor the floor zones can be decided until these duct sizes are known. In practice, preliminary design solutions, or estimated calculations, based on the experience of the leading designers, are often sufficient to resolve such matters. Such draft designs and estimates should be confirmed at the earliest possible opportunity, because many man-hours of work may have to be repeated if the assumptions prove to be significantly wrong.

The following are examples of high-risk nodes:

- At an early point in detail design work, one discipline, generally the architect, issues a set of key dimensions for the proposed building, which all the disciplines then use to co-ordinate their designs. These base drawings include plans and sections for every part of the building and its site. If the issue of this information is delayed, then the entire detail design work programme can be set back.
- Design information for the tendering of construction work must often be co-ordinated between several specialist designers, depending on the package. Any one discipline could delay the issue of this information and, in turn, delay the dependent construction operations and, potentially, delay the hand-over date of the building.

It is essential to identify high-risk nodes in advance, so that every effort can be made to reduce the criticality of the related activities. It often happens that high-risk nodes cannot be eliminated. In such cases, the plan does at least highlight the risk and focuses the minds of the design team on their responsibility for the related work.

4.12   DIFFERENT WAYS TO LINK DEPENDENT ACTIVITIES

In practice, few design tasks are heavily dependent upon the completion of preceding work. Design activities often start on the basis of preliminary information. This characteristic is very useful in reducing the criticality of particular design tasks, because it allows them to overlap.

This flexibility can be used to reconcile a design work programme with the time that is actually available for each stage of the work. Where designers agree to start on the basis of preliminary information, they are, in effect, dividing the task into two or more stages: preliminary work and final work. In this way, they are anticipating the need to iterate the design tasks. This enables the planner to refine the programme on the basis of shorter, more flexible activities.

Activities that are divided in this way can be represented as separate sub-activities, or the complete activities can be shown as linked in different ways. Figure 4.6 shows the various ways that such links can be formed. Some network planning software supports these different ways of linking activities and will rapidly calculate activity dates and float, taking these into account.

4.13   RESOURCE PLANNING

Network analysis of the design process for a project will suggest a minimum reasonable duration for the work, but it may not be possible to supply sufficient designers to meet the demands of this ideal timetable. To overcome this problem a

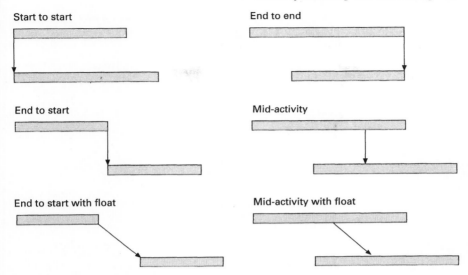

Fig. 4.6   Six ways activities can be linked in a network

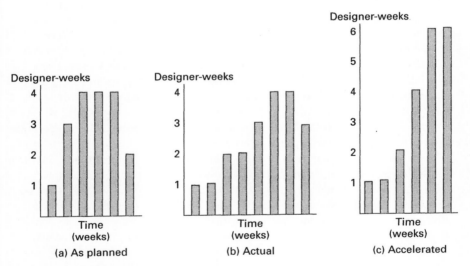

Fig. 4.7   Example resource histograms

technique known as resource planning is used. Network planning software generally incorporates a resource planning facility, which enables this to proceed interactively with activity scheduling, so that the resulting time and resource plans are compatible.

   Figure 4.7 shows three resource histograms, in each the vertical scale represents numbers of design staff and the horizontal scale represents the time in weeks, each

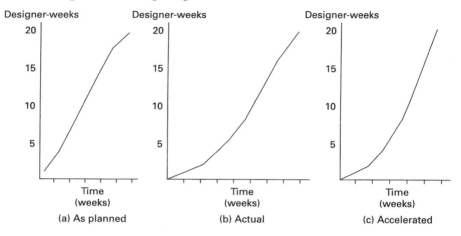

Fig. 4.8   Example cumulative staff utilisation curves (relating to histograms in Figure 4.7)

column being one month. Figure 4.7 (a) illustrates the monthly staff requirements needed to meet a particular planned programme of a fairly large-scale project. Figure 4.7 (b) demonstrates the effect of having fewer staff than the optimum in the early months of the project, based on an analysis of the work. As can be seen, the time-scale has been extended by two months. Had the problem been recognised in good time, it might have been possible to maintain the desired completion date by drafting in the additional numbers of staff as shown in Figure 4.7 (c). The availability of extra staff could depend on a number of factors, such as the size of the design practice, how many projects are running at the time, whether the project is of a common type or requires staff with special training and experience, whether temporary staff can be obtained from an agency, or if arrangements have been made to share work overloads with other practices.

The consumption of a resource can be plotted on a graph against time. The three graphs shown in Figure 4.8 show the same hours worked by staff as the three histograms shown in Figure 4.7, but as cumulative curves. In each illustration, the total weeks worked is the same, twenty. Figure 4.8 (c), the accelerated curve, is quite common in practice and this arrangement is often the most productive. It is often best if a small number of designers are allocated to a project at the outset, to analyse the design problems and study the general options, in liaison with the leading designers in other design disciplines. This prepares these designers to guide and integrate the work of a larger team, which is formed to complete the work by its appointed deadline.

4.14 COMMUNICATING THE WORK PROGRAMME

Whereas the leading designers should be consulted about the details of work planning, the project master plan is often dictated by the client. The client may, in this case, be acting on advice from a key consultant who is independent of the design team. In these situations, the design companies may be obliged to accept and work to a predetermined schedule of milestones.

Excessive detail in a programme can be confusing and makes it liable to get rapidly out of date. If a programme does not bear a clear relationship to what is happening, it becomes difficult to understand and impossible to work to. However, if the details of a programme can not be shown, it is quite important to ensure that the leading designers are properly apprised of the logic of the design production sequence. Programme briefings give an opportunity for the design team leaders to comment on the proposed programme and they are quite likely to see ways in which it could be improved. If the design team leaders are permitted, in this way, to give some input to the programme, it also becomes more likely that they will 'buy in' to it and, hence, be motivated to ensure that work actually proceeds according to the agreed programme. This level of involvement also ensures that they are able to make the priorities, implied by the logical sequence of work, clear to design staff. Design staff will then also appreciate the purpose and value of the programme.

Clear, simple planning charts, that have been co-ordinated between the disciplines, should be issued to each section of the design team. Ideally, these should be pinned up on a wall where they will remain in the view of all the designers. These charts can be marked up to show progress and small amendments that may be agreed from time to time. This practice of 'advertising' the programme can also assist managers to gauge progress if they are visiting sections of the design team, since the planning charts may direct attention to the most important design work in progress, or recently completed.

Each team leader has to convert the programme into a task list for the individual designers of the team, generally on a week-by-week basis. This list may look something like the example shown in Figure 4.9. In a larger team, the task list may be worked out together with one or two senior designers, who would then pass on instructions to the individuals who should do the work. In smaller teams, the leader would communicate the list of tasks directly to the designers, most likely at a short weekly meeting.

4.15 MONITORING PROGRESS

The purpose of monitoring is to ensure that work done, by any particular date, matches or surpasses that anticipated by the planned programme of work. If work satisfies the requirements of the programme, then the managers and the client can feel reasonably confident that the next milestone will be achieved on time. If work is not proceeding according to the plan, adjustments have to be made in one or more areas, which include the staffing, the programme, the effort that staff are putting into their work and the quality of liaison between designers and specialist design teams.

X - Y - Z ARCHITECTS		Team: B			TASK LIST for week ending: 27/2/1998
PERSON/ PROJECT	WORK STAGE	ACTIVITY/ DRAWING	TARGET HOURS	DAY/DATE REQUIRED	COMMENT
ANDY					
Hotel	B	Update layout	7	Thursday	CAD
Hotel	B	Elevations	7	Thursday	CAD
Hotel	B	Perspective	7	Thursday	CAD
Office	E	Foyer layout	14	mid March	Study
JAMES					
Office	D	Planning grid	14	Wednesday	Hard copy
		Group check of Planning Grid	1	Thursday	Meeting
Office	D	Wall/roof sections	10	next week	Insulation study
Office	D	Toilet layouts	10	next week	Check against services schematics
MARY	etc. ...				

Fig. 4.9   An example task list

Whereas planning is done from the top down, monitoring proceeds from the bottom up, by collating and summarising progress on the large number of activities that might be in hand at any one moment. For the measurement of progress to be reasonably accurate, the activity plan should include sufficient markers to compare with actual work, as it proceeds. Milestones are insufficient, because these are spaced too far apart. In design work, many tasks, or activities, may be in progress at the same time, so it is not sufficient to measure progress only by the tasks which are finished.

One way which is used to assess progress, at regular intervals, is to estimate the percentage complete of each scheduled task and aggregate these into an overall view of progress for the project. This is illustrated in Figure 4.10, which suggests how progress might be reported in the fragment of the office design project shown in Figure 4.5 (page 42). The week numbers shown for the 'actual start' and the 'actual finish' accord with the revised plan shown in Figure 4.11. This has been done, to show how the same progress can appear, when represented as a Gantt chart (Figure 4.11) or as a schedule (Figure 4.10). The result is that Figure 4.10 shows progress to be ahead of the base plan. This favourable circumstance is quite possible and one way to achieve it is discussed in section 4.18 (page 55).

Progress in design work has often been measured as the percentage completeness of drawings that will comprise the output of the design project. However, this approach does not necessarily yield an accurate view of overall progress, because activities in the list vary in duration and in the quantity of resources they consume. To produce a correct view of progress, these factors must also to be taken into account, together with others, such as the degree of co-ordination between developing designs.

When the overall percentage progress has been calculated, this has to be compared with that implied by the base plan, to ascertain how far the current

Progress End of week 4	Scheduled week no.		Actual week no.		Per cent complete (%)
	Start	Finish	Start	Finish	
STRUCTURAL ENGINEER					
1.1 Design floor system	2	3	2	3	100
1.2 Outline beam profiles	2	3	(task eliminated)		n/a
1.3 Stair and toilet core	9	10	5	–	25
1.4 Plantroom floor and enclosure	10	11	unstarted		0
ARCHITECT					
2.1 Planning grid	1	1	1	1	100
2.2 Toilet layouts	4	5	2	3	100
2.3 Wall/roof sections	1	2	1	2	100
2.4 Ceiling/floor zone	6	6	4	4	100
2.5 Stair sections	7	8	4	–	25
SERVICES ENGINEER					
3.1 Heating/cooling loads	3	3	2	3	100
3.2 Duct and core schematics	3	4	2	2	100
3.3 Duct sizes	5	6	2	3	100
3.4 Locate plant and tanks	2	2	1	1	100
3.5 Plantroom size and layout	7	9	4	–	50

Fig. 4.10   Design activity progress estimate

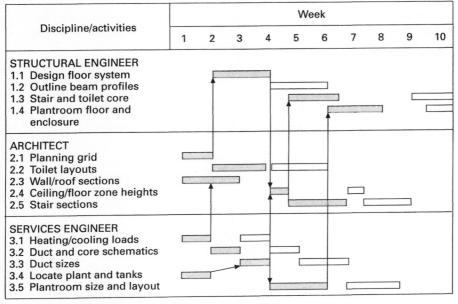

Key: ☐ original programme
         ▨ revised programme

Fig. 4.11   Revised design programme (developed from Figure 4.5)

situation is ahead or behind that required to complete the work stage by its assigned milestone. It can be helpful if a cumulative curve, indicating the anticipated 'value of work done', is prepared along with the base plans for each stage of work, for quick comparison with measurements of progress as work proceeds. Many proprietary network computer programmes can produce such curves automatically, comparing the base plan with progress data which is input as the work proceeds.

4.16   REPORTING PROGRESS

In any organisation, each level of management is, naturally, concerned that the work for which they are responsible should progress as smoothly as possible. To make sure of this, most managers go to see how the work is progressing, for themselves. However, in all but the smallest of design practices, senior partners are not involved closely enough with design production to make an accurate assessment of the overall progress of particular projects. In general, they only need to know if problems are arising, who is to solve them and how they are going to go about it. A formal procedure and communication structure has to be set up to feed these managers with the information about progress that they and their clients need.

Part of the work of senior designers is therefore to gather progress information and summarise it for presentation to managers. The senior designers generally report on progress to their superiors more frequently than senior managers would make similar reports to the partners of the design practice and their clients, so that any small deviations from the plan can be corrected immediately by staff. The senior managers need be involved only when staff or other resources need to be expanded or reallocated. Such action can only be determined by the senior management since it can impact on the profitability of the work, or the overall performance of the project and, possibly, the progress of other work in the practice.

With this structure, the decisions that are needed to maintain progress on a design project are taken at the level of the manager or supervisor, according to their capacity to control staff and directly monitor the success of remedial actions.

4.17   ACCELERATING WORK

In principle, where work is behind schedule, output can be brought back to a previously agreed programme by adjusting production rates. This generally requires a commensurate adjustment to be made to the quantity of input to a transformation process. In practice, there are often circumstances which make it difficult to accelerate the work of a design office. If additional staff are brought in, they must first spend time becoming familiar with the acceptance criteria and design work done so far. The productivity of the established members of the design team may also diminish, as they spend time instructing the additional staff about the work. It can also be difficult to co-ordinate design work that has to be done simultaneously, that was originally planned to take place sequentially.

Thus, accelerated working generally costs additional designer-hours. Not only does it tend to be expensive, but it can also be stressful for people working in the design team. One of the main benefits of planning and monitoring design work is to avoid such 'crisis management' by maintaining optimal pressure for production, taking into account the personalities of individual designers.

4.18 PLANNING NETWORKS DEPEND ON THE DESIGN

Opportunities to improve the efficiency of design production can be found in the design itself.

The operational plan for design work can sometimes be improved by adopting particular design options. To accelerate the plan shown in Figure 4.5 (page 42), time has to be saved somewhere on the critical path. This can be done by making the development of duct and core schematics (activity 3.2) independent of the wall and roof design. Target insulation values could be agreed in advance by the architect and the heating and ventilating engineer. The architect would then design the walls and roof accordingly (activity 2.3). This saves two weeks on that particular path, but this puts the floor system and beam profile design work onto the critical path (activities 1.1 and 1.2). The designers could then opt to use a flat floor slab, to eliminate the need to 'outline beam profiles' (activity 1.2).

The revised plan is shown in Figure 4.11 (page 53). The revised timing of tasks is shown by the hatched bars, while the original programme is shown by the unhatched bars. In Figure 4.11, nearly all the activities are on a critical path. Not only is there more than one critical path, but each of these paths has a margin of float. This characterises a good plan: the activities are closely stitched together and staff are kept on their toes by the close liaison that this necessitates, but there is, nevertheless, more than sufficient time to get the work done and little cause for anxiety.

To gain time by changing the design requires close attention to the design work programme early in the design process. With experience, designers become aware of the forms of construction that are quick to design and of design configurations that would require more time to complete. Of course, only design options that would meet the requirements of the brief could be considered.

If a programme is devised that allows an earlier completion than originally planned, the managers and the client must decide whether to bring forward the target deadlines, or use the margin gained to ensure a better quality of design work. The latter approach could simplify construction and provide it with better design information, which might, in turn, reduce the construction cost and bring forward the final hand-over of the building. More time spent on the design could also enhance the value of the finished building and bring continuing dividends through reduced running costs and sustained asset value.

It can be seen that careful production planning can increase the range of choice available to the client and improve the opportunities for surpassing a project's objectives.

4.19   EFFICIENCY, PRODUCTIVITY, LEARNING AND ADDED VALUE

The efficiency of a process is a measure of how much is lost along the way. The waste, as shown escaping from the transformation process in Figure 2.1 (page 12), is a loss of potential production or profit. Rather than considering output minus input, efficiency is generally measured as output divided by input. If numerical measurements of input and output are available, this can be expressed as a percentage. The problem in measuring efficiency is that, generally, inputs are in a different form from the output. If financial values are attributed to these entities, it can be seen that design work must give 'added value' to its client to justify investment in the consultants' fees. This reaffirms, somewhat obliquely, while production has to change the state of things, from one that exists to one that is more desirable, design work is concerned primarily with achieving an *outcome* that satisfies the client, rather than short-term efficiency for the design practice.

In design work, a lot of time may appear to be wasted and paper put in the bin. Design development processes look very inefficient, as many of the ideas first put forward are set aside or amended beyond recognition. Occasionally, unsuitable design ideas can be salvaged and put to good use in later projects. Thus, a lot of design effort goes into a process of continuous learning, which develops the knowledge, experience and expertise of the designers throughout their careers.

Computer aided design has undoubtedly increased productivity, in terms of how long it takes designers to do their work. It typifies how greater productivity is obtained, at the cost of the time it takes to learn the necessary skills, and build up libraries of design elements that can be reused. Once this training has been done and the library built up, the full advantages of CAD can then be implemented in all future projects.

In general, waste and opportunities for learning are considered as overheads to design work, which are not shown in design work programmes. However, there are often occasions where staff must obtain special knowledge for particular projects. Such learning takes time and it may be recognised as specific input to design development. It may therefore be appropriate to show some specific learning activities on design work plans, complete with inputs (such as a briefing and staff time), a duration and outputs that connect elsewhere with the design process. The briefing given to design staff, when they join a project team, could be included in this category of activities.

SUMMARY

Planning generally proceeds from the top down. The main stages which are identified in the project master plan are successively analysed in greater detail.

Experience is needed to define design activities and their relationships, in order to establish a precedence network. Advice from all the leading designers in the team is generally needed and a lot of imagination may also be required.

In design work, iteration and back-tracking can easily eat up available float and convert non-critical paths to critical ones. Design work planners should be able to recognise high-risk nodes and any links to activities and deadlines that lie beyond the next milestone.

The design teams should be fully aware of the work plan and appreciate how their output will be used by other designers and in the construction operations. If this is the case, planning should enhance co-ordination between different sections of the design team. Adherence to the programme may be helped if problems and proposals are communicated clearly and discussed at the appropriate time. In some instances, the programme of work may be improved by adjusting certain characteristics of the design itself. The success of planning is therefore dependent on forward thinking and effective communication.

Monitoring progresses from the production level of the design team, up through the management hierarchy. The design team leaders should be familiar with the programme and understand the planning and monitoring techniques in use, for this reporting process to work effectively. Their knowledge should include how to estimate the completeness of design tasks and an appreciation of how these are aggregated and summarised in progress reports to senior management and the client.

Accelerated working may create float in a planned way. Unplanned acceleration reveals poor planning and monitoring and leads to errors and inefficiencies. A good programme probably includes many parallel critical paths, each with a margin of float.

NOTE

1. The technique of network analysis is explained, for example, in *Construction Management in Practice* by Richard Fellows, David Langford, Robert Newcombe and Sidney Urry, Longman Scientific and Technical, 1983. Network analysis is described and discussed in relation to other planning techniques in *Modern Construction Management* by Frank Harris and Ronald McCaffer, Granada 1977, second edition 1983.

CHAPTER 5

FEASIBILITY AND OUTLINE DESIGN

5.1 INTRODUCTION

People embark on a building project because they believe that new, or refurbished, accommodation will improve their lives or businesses. The transformation of such vague ideas into very precise instructions to a builder, about what should be constructed, not only requires design skill, but also organisational ability. This chapter focuses on the first things that have to be done to realise a client's vision of the future.

The proverb 'well begun is half done' stands the test of time. It can be difficult to begin design projects well, because there are usually many uncertainties at this stage. The site and sources of finance may not have been identified and the client may not be fully aware of the stages of design work, the need for statutory approvals and financial risks that attend construction projects.

Initial design work requires little formal planning, partly because these uncertainties may render this somewhat futile, but also because few people are involved in project inception and their work is relatively easy to co-ordinate. Nevertheless, preparation for work planning at this stage can be vital to the generation of a workable design work programme later on. This process begins by gathering sufficient information into the project brief and developing the quality plan, to ensure that subsequent design work is approached systematically.

The second part of this chapter discusses differences in the outlook of various participants in a building project. This is significant to planning and monitoring work for several reasons, not least, because each group will have its own view of what most needs to be controlled. The necessary process control systems should be initiated as soon as practically possible. The chapter concludes with a view of the work breakdown that includes these control mechanisms, to assist in putting them into operation. There is also a brief look at the need to register outstanding information and identify risk factors that could slow progress.

5.2   THE MAIN DELIVERABLES OF INITIAL DESIGN WORK

The first steps of project initiation require little formal planning, because projects are generally defined by an executive team that works in close collaboration. Similarly, initial design concepts are usually generated by a small group, where individuals work closely with one another. Nevertheless, a simple breakdown of the required deliverables and a time schedule for their production allows the lead consultant to keep track of progress and communicate more effectively with the client.

A breakdown of early design work as it might be seen in common practice is shown in the linked bar chart illustrated in Figure 5.1. This represents an actual project, where the options and concepts for spacial, structural and building services systems were developed in an integrated way, requiring input from a multi-skilled design team.[1] Each bar represents a set of activities that would result in an output, or deliverable, varying in form from a letter of appointment to sets of calculations and presentation drawings.

The principal activities during the early phases of a project are to develop the design brief, generate a concept design which satisfies the client and his financial sponsors and obtain the necessary statutory approvals for the development to take place. Some research into the particular requirements of the building and the site (or alternative sites) often has to be carried out, as well as exploring various design options. The search for finance is an activity which often delays the start of a project.

In Britain, approval for building development is required from local authorities in all but exceptional cases. Traffic planning departments often have control over the accessibility of a site. Since this can be a big influence on the layout of a building, their advice may be relevant at an early stage of design work. Approvals may be needed from English Heritage, or Scottish Heritage, if a building of historical interest is to be altered or if the site of the proposed development is in a Conservation Area. Some building uses may need additional permissions from the local authority, such as licences for gambling and the sale of alcohol. Very occasionally the Health and Safety Executive may have a decisive influence, for example, where development is proposed in the vicinity of a gas pipe-line that could pose an explosion risk. In general, the local planning authority will advise upon the need to consult other statutory bodies, but clients may be wise to engage a planning consultant (or development surveyor), to ensure that no significant matter is overlooked.

All being well, the outline design phase delivers a design to the client, which incorporates all the relevant advice, has received development approval and can be taken forward to the detailed design stages, in preparation for tendering and construction.

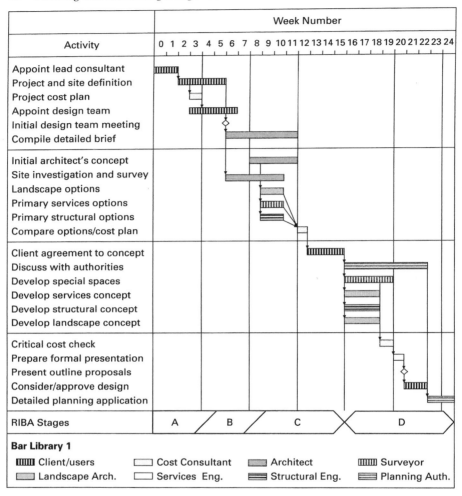

Activity	Week Number
	0 1 2 3 4 5 6 7 8 9 10 11 12 13 14 15 16 17 18 19 20 21 22 23 24
Appoint lead consultant	
Project and site definition	
Project cost plan	
Appoint design team	
Initial design team meeting	
Compile detailed brief	
Initial architect's concept	
Site investigation and survey	
Landscape options	
Primary services options	
Primary structural options	
Compare options/cost plan	
Client agreement to concept	
Discuss with authorities	
Develop special spaces	
Develop services concept	
Develop structural concept	
Develop landscape concept	
Critical cost check	
Prepare formal presentation	
Present outline proposals	
Consider/approve design	
Detailed planning application	
RIBA Stages	A　　B　　　C　　　　　D

Bar Library 1

Client/users	Cost Consultant	Architect	Surveyor
Landscape Arch.	Services Eng.	Structural Eng.	Planning Auth.

Fig. 5.1   Outline design programme

5.3   EARLY DESIGN PHASES

Before detailed design work for construction can proceed, the design and the organisation of the project require very careful forethought. This preliminary work can be divided into the three main stages, as introduced in section 3.4 (page 25). Their purpose and content are outlined below and then described in more detail in the sections that follow.

1. *Project definition*: When a building project is initiated, its aims are examined, to ensure that they are self-consistent, that building work would address the client's

needs and that construction would lead to a return on the investment of finance. Professional advisers may introduce inexperienced clients to similar projects that have already been completed, to help them to agree reasonable expectations of what the project should achieve. In addition, conceptual sketches for the proposed building can be offered to the client.

Development of an outline design brief, and exploration of the particular opportunities offered by alternative sites, will occupy a period of time dependent on the complexity of the issues. Almost any scheme can be built; the important question is whether or not the proposed location and form of the building would meet the actual needs of the client. For this reason, estimates of likely building costs are needed right from the beginning of a project.

If matters of public interest are in question, the feasibility of a project can be a political matter, with unpredictable effects to the timing, funding and definition of the project. Where any uncertainty applies, such as public controversy about a proposed development, research into design limitations and options can be of unpredictable duration, as can the related consultations and possible planning appeals. It can be difficult to identify in advance every activity that should be done in such circumstances.

For all these reasons, this feasibility stage may overlap substantially with design development, especially if the client wishes to press forward to obtain official approval for the proposed development, even before the detailed requirements of the design have been agreed. This point is illustrated in Figure 4.1 (page 37), where the site is not actually acquired until design work is well advanced.

2. *Development of outline design proposals*: Conceptual designs are generated along with the outline brief, to crystallise various aspects of the proposals and to ensure compatibility between the proposed spacial, structural and building services systems. This stage of design development includes attention to architectural concepts, which set the tone of the proposed building or buildings (prestigious, conservative, innovative or otherwise).

Predictions of the likely cost and time–scale of construction of an outline design can be influenced by factors such as the availability of the vehicular access and the mains services to the proposed site. Together with statutory approval for development, such matters must be resolved before the client would give consent to proceed to more detailed design work and, in some cases, before the site is bought.

3. *Production of the scheme design*: If the outline design and estimates of the implied cost are acceptable to the client, the financial backers and the local authority, then these proposals are developed in more detail, to seek full approval for the development from the local planning authority. The scheme design also provides the level of detail, about how the building is laid out and what it contains, that is needed before the detailed design work can begin.

This work may proceed while the detailed terms for securing title to the site are being negotiated.

Each of these stages is needed to ensure that the detailed design brief is properly developed and to provide time for initial design ideas to mature. It is not unusual for building layouts, structural systems and similar fundamental considerations to be restructured once or twice during this process, in a more or less comprehensive way. It is far better that this should occur during the conceptual development, rather than later on, when improvements can only be bought at the price of reworking substantial amounts of design information.

Under continuing pressures for efficiency, there is a temptation for clients and design consultants to reduce the effort given to these preliminary stages. In all but the simplest design work, this would be ill-advised, since deficiencies would have to be made good later on, at the cost of additional expenditure and delay. If early design work is hurried, and no resources are available to improve it later, it is unlikely that the resulting building will fully satisfy the needs and preferences of its owner and users. Some repetitive building types may be an exception to this generalisation, not because their designs do not need careful preparation, but because this has been done, in effect, through a long series of similar projects that have already been carried out.

5.4   PROJECT DEFINITION

The essential task of project definition is to clearly describe the main purpose of the project and the various constraints which apply, including the time and financial limits. A formal project definition statement may be written, for communication to all the design consultants and other people who need to understand the aims of the work, including the financial backers of the project and staff in the client organisation.

The project definition statement may include an explanation of how the aims and objectives of a project accord with the client organisation's policies and business (or other) strategies. Usually, the over-riding purpose, from the client's point of view, is not to build a building, but something else. For example, a business may seek profit from a new production unit, or a local authority might require a facility that brings visitors to an area.

Figure 5.1 (page 60) shows a typical programme of work for the project definition, outline design and scheme design stages of a project. It should be kept in mind that work on the brief and the designs normally proceeds in overlapping phases, as discussed in section 3.3 (page 24), so that, in practice, the stages are defined in order to set targets during a period which is often characterised by uncertainty. In the example in Figure 5.1 (page 60), preliminary work proceeds for three months before the feasibility of the project is established, as indicated by the bar labelled 'client agreement to concept'. This agreement followed the outcome of the analysis of the cost of construction of a particular outline design, as shown by the bar immediately above: 'compare options/cost plan'.

The early stages of a project offer opportunities to challenge the project definition statement, the outline brief, the proposed arrangements for organising design and construction work. This managerial work would include attention to the contracts that are to be set up for design and construction work.

5.5 BRIEFING THROUGHOUT THE PROJECT

Several kinds of brief are used in building projects, as shown below. Broadly, their function corresponds to the calibration of production tools referred to in section 2.8 (page 17). Each has to be programmed into the project timetable so that required information is available when it is needed.

1. *The project definition*: as described in section 5.4, above.
2. *Instructions to consultants*: These set out what contributions are expected by the client from each consultant. The lead consultant usually discusses these contributions with candidate consultants, to clarify how they will collaborate to produce the required deliverables.
3. *The outline design brief*: This is a preliminary interpretation of the client's statement of project aims and objectives. Its purpose is to enable initial research to be conducted into the building type and design options, leading to the generation of conceptual designs. This brief sets out, in general terms, the purposes and scope of the project, together with known constraints including the construction budget and target running costs. It would normally include a schedule of required accommodation, with an analysis of the main activities that are to be housed. This analysis helps to establish the relationship between elements of the accommodation and how they interact.
4. *The detailed design brief*: This sets out the requirements that must be satisfied by the elements of the building. It may include detailed information about the activities that a building will accommodate, for example, the requirements for the strength of the floor, building services, daylight, furnishing and fittings, wall, floor and ceiling finishes, characteristics of the site, constraints imposed by the planning authority, preferred materials and forms of construction.
    The detailed design brief is usually worked out when concept designs are being developed into the outline and scheme designs. It is submitted to the client for acceptance as the blueprint for the detail design and then incorporated into the structure of the quality plan, used to control all further work.
5. *Team induction*: This is the process by which new members of the project organisation are introduced to the project, its objectives and all the other important information that they need. Such briefings may be verbal and visual. They should be delivered by established project participants. Briefings should also be adapted to ensure that they are particularly relevant to each team that joins the project, although the main core material must be consistent.
6. *The project handbook*: This document contains all the key information about the project and the participants in the project organisation. It should outline the aims of the project, its timetable and basic information about the design. It should include contact details of all participating companies and their

responsibilities, possibly illustrated by an organisation chart on larger projects. Key documents should be listed, together with an explanation of administrative systems, such as how the document management system should operate and how progress should be reported. The project handbook can explain key areas of the project quality plan, for example, as a key reference for initiating and formatting communications.

Since construction project organisations grow and change, the handbook cannot be considered to be a one-off document. To retain its usefulness, it should be regularly updated as design work proceeds. The receipt of updates should be acknowledged by each recipient, to ensure that every party to the project is working with the same handbook at all times.

Briefing does not finish at any particular point in the project. Every time a new company or individual is introduced to the project team, the main objectives need to be explained to them, to help them to understand the contribution that is expected from them. In most cases, briefing is not just an issuing of instructions, but a process of two-way communication. People are brought into a project because of their particular knowledge and skills. When they study the instructions they are given, they may seek answers to questions that others may not yet have put forward. Many specialist designers will bring with them sets of design requirements, some of which may impinge on work already being done by others.

5.6   DESIGN BRIEFS

Careful analysis of the project definition statement, and consideration of how other buildings may have satisfied similar project objectives, yields a set of design requirements. This information is collated to form a design brief. This brief may develop in definition and detail through several stages. It is usually necessary to create conceptual designs on the basis of a preliminary outline brief because design work is actually necessary to fill in the detailed aspects of the final brief.

It may sound paradoxical that design work should be done before the brief, but in practice, initial images of a proposed building, presented by the design team, can help clients and the potential users of a building to become more articulate about their aspirations and thereby help to distinguish between the client's declared wants (a 'wish list') and their actual needs (minimum acceptance criteria). Early design sketches also help designers to recognise the potential of a site and such vision can improve the possibilities for a return on investment and, thereby, ensure that the proposed development is viable.

The most intense briefing activity is, generally, early in a project, when the key objectives for design work are set. However, the fine detail of requirements may not be entirely known until detail design work begins, or even later. An architect often has the greatest input to this process, since brief-taking is a primary skill of that profession. A well-chosen architect will have extensive knowledge of the project type and buildings that have already been constructed (or refurbished) to satisfy similar requirements. In some cases, a development surveyor may be appointed to advise on the commercial aspects of the outline brief. For many projects, service engineers also have a very important role to play in preparing the brief, as building services may consume as much as 30 per cent of a construction budget and dispose of an even higher proportion of the life-cycle running costs of a building. For example, where a choice must be made between natural ventilation and air-conditioning, expert knowledge of the technicalities would usually be needed.

Standard briefs can be used where an organisation regularly procures similar buildings (for example, houses, offices and schools). However, these should leave scope for reconsideration and modification in the light of special circumstances, such as the preferences of a client, the unique qualities of a site and the space required for telecommunications cabling. Although design teams generally hold the primary responsibility for taking account of developments in building technology, legislation, and so on, a significant research effort may also be undertaken by client organisations, to provide information about what it should accommodate, how it will be operated and, often, the appropriate image that the building should present to occupants, visitors and the public.

Research into such requirements must sometimes be dove-tailed into a client's own programme of business development. Large numbers of staff may be consulted, requiring the organisation of opinion surveys, analysis of business practices and so forth. In some cases, such as leisure, housing and retail developments, questions may be put to a sample of potential customers. As a result, the early efforts of the design team may be surpassed by personnel on the client side, where considerable organisational and diplomatic skills may be employed. Alternatively, specialised research companies may be employed. Work on this scale would not commence until the feasibility of a project has been sufficiently validated to guarantee a return on the effort and expenditure. Also, where large numbers of people are involved in gathering information, this inevitably communicates a great deal about the importance that the client has assigned to the project. This can have an impact on the morale of staff in the client organisation, for instance, because a physical move for operations often implies changes in the client organisation structure and job definition and availability. New buildings are also likely to reflect the client's public image, so such investigations may also have to done in liaison with senior marketing managers, in addition to the involvement of operational and personnel managers.

5.7   PROCESSES OF INITIAL DESIGN PRODUCTION

Figure 5.2 clarifies the relationships between the main processes, described above, that lead to the production of a scheme design. The time-scale goes from left to right.

The statement of project objectives is the starting point and this continues to stand as a primary reference for all the design work, as indicated by the arrows that flow from it. The design brief is developed from this statement and sets out the

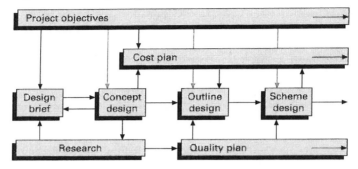

Fig. 5.2   Initial design activity

main requirements of the building or buildings, as shown by the flow of arrows from left to right. The arrow pointing left from the concept design recognises that development of the design brief is often assisted by the ideas and questions that emerge from this work. Concept sketches help to clarify the ideas about such matters as the relationship between the building in the proposed neighbourhood, its principal spaces and overall shape, together with the main structural and services systems and the architectural style.

Research, while the initial conceptual designs are being generated, would include investigation of the site, examination of existing buildings of a similar nature and close consideration of the activities that the building should accommodate. Some attention may also be needed to ensure that statutory considerations are properly understood by the design team, including constraints on development that may be imposed by the authorities. Relevant information gathered by this research may be included in either the outline or the detailed design brief, which is a key reference in the quality plan. Research tends to continue throughout all the design stages, because new questions arise as each level of detail is explored. As far as possible, these tasks should be represented on the work programme, to show how their output feeds design development and have appropriate priority assigned to them. Such detail would be hammocked in the detailed networks represented by the brief development bar on the high-level planning chart.

The production of conceptual design sketches affords the earliest opportunity for quantity surveyors to analyse how much a design would cost to construct, as indicated by the arrow connecting these boxes in Figure 5.2. The cost plan incorporates the budget allowed by the project objectives and is progressively refined in detail and accuracy, throughout the development of the design. It may begin as an initial rough estimate, to check feasibility, followed by some research into the construction costs and the running expenses of similar buildings. This information, together with an analysis of the initial concept sketches, enables an initial estimate of the probable cost to be made. As the design evolves through the outline and scheme design stages, the cost plan develops interactively with it, because the cost is a very important constraint on key decisions, such as how big the building will be, how it is to be constructed and what it will contain, for example, in terms of the services systems, finishes and equipment.

The quality plan should describe the procedures by which the design brief and conceptual design will be developed into an integrated design, because any deviation from a logical sequence of design development could lead to inadequacies in the quality of the design output. When set against a time-scale, this sequence of activities and products becomes the programme for the design work. The availability of senior designers also has to be taken into account, so that they will contribute to the development of the design in key areas and at critical points in the sequence of the work.

Figure 5.2 is not set out as a bar chart, to make the point that the processes it shows may not have clear-cut beginnings and ends. Information about quality acceptance criteria may continue to raise questions about options and priorities, at different levels of detail, throughout the design process. The quality plan itself should be further developed, so that it is fully adapted to the purpose of each phase of design work when it begins.

5.8 PLANNING OUTLINE DESIGN WORK

While the stages of design development are common to the production of almost every building design, the detail of the tasks may vary enormously according to the size and complexity of the project. Experience with similar clients and projects helps to determine the essential steps and likely durations of early design work. Figure 5.1 (page 60) is abstracted from an actual design process, where the development team was close-knit, the concept was accepted immediately and development of the outline proposal was straightforward. By contrast, proposals for some projects have been known to pass between the designers, clients and their advisers for years, before a conclusion was reached.[2]

Experience of the project type, location and changing market and economic environment is often indispensable to the creation of realistic time plans for outline design work. The pattern of communication required to develop an outline design and conclude feasibility studies is also unpredictable. There are occasions where a detailed planning chart can assist discussions about how to co-ordinate short bursts of activity, such as presentations of research findings or design proposals, but it is difficult to prescribe general rules for these situations.

Visual and other aspects of building design may require unexpected levels of effort. If the site is a sensitive one, for example a historic location where only limited development would be permitted, consultations may be needed with authorities such as, in the case of the historic site, English Heritage. This can lead to amendments being imposed as a condition of granting planning approval. At other times, work on the outline or scheme design can be extended because the client seeks an innovative solution, for example to achieve a certain prestige, or to accommodate innovative processes or services.

Expert help and advice are frequently sought, both in generating designs and in reviewing them. The timing and extent of this input from design staff, such as graphic artists, consultants and urban designers, does not always have to be agreed in advance, depending on the amount of time that is needed, but work planning charts can be useful to inform potential participants about the earliest and latest dates when their input is likely to be needed. Forethought about what advice to seek may help to minimise the occurrence of iteration and back-tracking during preliminary design work, as it does for detailed design work.

5.9   CHANGE CONTROL

Substantial areas of the detail design brief may, deliberately, be left undefined until quite late in design development, especially where technology, required for the building or to be accommodated in it, is advancing rapidly in relation to the project programme. This could affect, for example, the layout of equipment in an industrial building, supermarket check-outs and baggage handling areas of airports. It can be particularly difficult to compile a brief for hospitals, because the equipment used and the methods of treatment employed may develop more swiftly than the building design and construction. The shortening of hospital stays and the speed of treatment has revolutionised the space required by some departments for treatment, staff and administration, in less time than it took to obtain planning approvals and prepare design information for construction. In such projects, latitude has to be given in the design to accommodate such changes.

There can be commercial reasons for leaving aspects of a design brief incomplete. For instance, it may be difficult to determine precise servicing arrangements for parts of a building where the tenants have not yet been identified. Sanitary accommodation is a case in point, for the detail design may depend not only on the number of people who may be working or visiting a given area, but also on their status as managers, workers, customers or clients. It is very important to identify every vague area in each brief and include target dates in the programme of work, when clarification of each point is required. This is to ensure that sufficient time will be allowed to trace all the knock-on effects of late decisions on other areas of the design, which must be worked through before information can be released for tendering and construction.

Designing from an incomplete brief requires a lot of skill and experience, including an ability to make decisions in an atmosphere of uncertainty. This is because, when the postponed decisions are finally made, the repercussions on the design work that has already been done can be substantial. Designing from an incomplete brief requires a strategic approach that takes into account the full range of possible decisions that might be made later. It also requires a total appreciation of all contingent systems of the design, the design information database (including drawings, calculations and specifications) and the construction process, so that as briefing decisions are firmed-up, every possible repercussion is recognised and addressed before the affected areas of construction are put in hand.

The discipline applied to programming late decisions and tracking their design and cost implications is referred to as change control. It is not always easy to distinguish changes that need such specific control from general design development, because this is also a transformation process. To exercise such control not only requires a detailed knowledge of the design and design development work in general, but also authority to assign staff to work on design changes and to block changes that could seriously delay other work or be unduly expensive. In many cases, the necessary knowledge of a particular project can only be built up through involvement from a very early stage of the work, by someone with a network of contacts in the client organisation. Further, to implement such control in every specialised design office that is working on a project, authority must be vested at the highest level. Thus, change control is generally managed by the lead consultant or the project manager, because they have this authority and direct access to decision makers in the client organisation. However, self-discipline is also needed in each of the design disciplines, to avoid unnecessary changes to partly completed designs.

5.10 THE PARTIES INVOLVED

The breadth of the project organisation, in terms of the number of companies, authorities and individuals concerned, and its height in terms of the managerial hierarchy, are significant to planning and monitoring for several reasons, not least that each group will have its own priorities and each level will be primarily concerned with different aspects of planning the job. This will influence the setting-up and effectiveness of control mechanisms, including those that govern the progress of design work. The establishment of these controls can be a complicated activity that, itself, requires planning, quality controls and monitoring in operation.

5.10.1 The project coalition

The individuals and teams who collaborate to mastermind, design and construct, or to refurbish a building, generally come from a number of discrete companies and organisations. The project organisation is, therefore, by nature, a temporary coalition of interests.[3] The procedures for building-up, operating and disbanding this organisation require considerable skill in design and operation, if everyone is to contribute effectively and the majority are to get the rewards they seek from their involvement.

5.10.2 The client

In the context of this book, the term client is used to refer to an organisation or individual wanting a building constructed or refurbished. Clients vary enormously. Some are exceptionally well organised and their construction projects are elements of a long-term business, or administrative, strategy, and each one is managed systematically. Such clients ensure that the brief is well worked out and includes terms of appointment for the design consultants, giving a very clear set of objectives and priorities. Their project teams are well constructed and communications orderly. In such cases, the person with overall responsibility for a project is often a member of the client's own staff. Other clients may have only a vague notion of what they want. Their first ideas are prompted by familiar buildings, which may bear little relationship to their real needs. Such clients may have no knowledge about how to organise design and construction work.

There are also 'client partnerships'. These vary in formality. Some projects may be directed by joint-venture companies specifically set up for the purpose. In other cases, the client consortium may be informal, for example, where the viability of a project depends on the commitment of a leading tenant. This tenant may have little direct control over the work, but nevertheless is consulted routinely about every decision that might concern them.

While experienced clients determine the objectives of a project for themselves and exercise control over its progress, either directly or through experienced agents, inexperienced clients are generally advised to direct building projects through an agent. For some building schemes, this may be an architect, but on large undertakings, a project manager is generally appointed. Project management services are usually commissioned from a company with specialised staff who are experienced in this field. This is often a quantity surveying practice, because this profession is acknowledged as having a superior grasp of the cost variables.

There are a limited number of professionals who specialise in giving independent advice to construction clients. Where these take no further part in directing the project, they may be referred to as the client's key adviser.

5.10.3 Stake-holders

The actions and decisions of a building client are often influenced or constrained by several groups which may play no active part in the project whatsoever. Together with the client, these people are collectively referred to as stake-holders.

The term stake-holder can be used in different ways. In most practical situations, those who manage construction projects need only take account of groups that have some direct power to influence the progress of the project. These stake-holders include the client's financial backers and may sometimes include important customers, such as tenants or purchasers of the completed works. Statutory bodies and utilities with power to influence designs and construction practices are also stake-holders in the project, as representatives of the public.

Town planning legislation allows adjoining neighbours to object to development proposals on a site and thereby influence the progress of the project. People in the locality also have powers, through environmental protection law, to obstruct construction work if they are inconvenienced by noise or other pollution.

The term stake-holder has also been used to include people who are influenced by a project, even when they have little or no influence over the decisions of the project managers. These stake-holders would include people who will use the completed building, organisations that will benefit from it, or who may, possibly, be adversely affected by it, for instance, if it houses a competing business. Also included would be the public in the locality of the building, who could be inconvenienced by the construction process and who may see the building as an enhancement or detrimental to the environment.

The environmental impact of a building can be even more widespread, for example, through the mining of materials from which it is made, or the use of fuel during the building's manufacture and subsequent use. Energy is incorporated in every building, through the production and transport of its materials and components. The heating and cooling of buildings may continue to generate greenhouse gases throughout its existence. It may seem somewhat exaggerated, but it is possible that people may therefore be a stake-holder in building projects taking place on the opposite side of the world, since these contribute to global warming and could damage their livelihood through consequent drought or flood.

Other issues, such as the use of timber from threatened forests, also generate distant stake-holders who may have indirect influence through political interests, or the activities of protest groups that can attract media attention. Some project types are especially susceptible to public concern, ranging from industrial factors that may threaten contamination, including airports that spread noise, to living units for people who may attract discrimination from the surrounding community because they are leaving various kinds of institution. Such potential influences on the progress of design and construction work should be taken into account by a risk management system, discussion of which is beyond the scope of this book.

It may also be relevant to note that every company or individual working, or seeking to work on a project, has an interest in its conduct and progress. These people are not normally considered to be stake-holders in a project, because their interests and influence are taken into account when contracts for their engagement are set up.

5.11 STRATEGIC, TACTICAL AND OPERATIONAL PLANNING

Besides the horizontal spread of the project coalition, there is also a vertical dimension to the project organisation. Board members, senior management, middle management, supervisors and operatives represent levels in the management hierarchy of their companies. These levels tend to have different primary concerns and to focus on different time horizons, with more or less concern for the planning of design work.[4] To understand this requires elaboration of the simple organisational breakdown shown in Figure 4.2 (page 39). The roman numerals given in each heading below relate to the recognised levels of a business organisation.

V   The board of directors, or senior practice partners, decide policies

Policies answer important questions, such as: 'why is a building to be built, or modified?', or, 'which business are we in?' While detail is avoided in policy statements, they may generalise about how a project is to be organised and how much time and money may be spent on it. For example, public property services were pushed by central government, in the 1990s, to award all contracts for work, including design services, to companies that tendered the lowest price.

Policies, generally, remain unchanged for many years. They do not usually alter during a particular construction project, but sometimes the senior managers of a client organisation, or a new sponsor (including a change of government), can vary them to suit a revised set of organisational objectives. One such policy change in The Netherlands revolutionised a whole programme of hospital building, by changing from a programme of building large hospitals, to providing local care by increasing the number of smaller ones.[5]

Company (or organisation) policies may be included in a project definition statement, where these clarify its objectives, or special constraints on it.

IV   Senior managers recommend strategies

In the military context, strategies are devised to gain advantage over an enemy, for example, by building a fleet of submarines that threaten nuclear retaliation from hidden and changing locations. Strategies tend to be implemented over long time-scales. In the Second World War, the Allies' decision to invade Northern France, supported by 'Mulberry Harbour', took three years to implement.[6]

In building projects, the selection of a site is generally a strategic choice, made by a client, often with advice from an expert, such as a development surveyor. For a design practice, the choice of which CAD system to buy is strategic, whereas the aim of keeping up with developments in information technology would be a policy.

Land use plans, prepared by local authorities, are strategic documents, that aim to ensure a balanced provision of housing, employment, transport and local space in any locality. Mixed developments, such as those that provide housing as well as shopping or business facilities, are frequently devised by developers as a strategy to satisfy their business objectives as well as the policies of a local authority.

Another strategic decision is the selection of a procurement method, for example, whether to use a traditional system, construction management, or design and build. Another may be whether to include air-conditioning in a building or put faith in an environmental engineer's capability to design adequate natural ventilation. Such decisions may influence which specialist designers are employed and the scope and the time-scale of their services. These decisions are generally taken by executive project managers and referred to the client for approval.

A key decision at the strategic level that often has to be made by the design team, is agreeing the equipment that should be installed to manage and communicate design information. On large projects, specialists in information technology may be appointed to advise and control this aspect of organising the work. Similarly, many companies may not opt for the best computer programmes, but buy those which are most used in their sector of an industry. In this way, they benefit from easy communication with other companies and find it easier to recruit staff who are already familiar with the software, without incurring the cost and delay of training them.

In a building project, strategic planning, including the allocation of finance, calendar time and staff, generally takes place several weeks before work begins on each of the major stages of design, construction and hand-over.

III   Middle management are concerned with tactical planning

This can involve a fair amount of calculation, for example, allocating the total time available to tasks and deciding which people are needed to do them. For design work in the construction industry, this level of control involves practice associates and design team leaders in architectural, engineering, cost control and specialised disciplines, with the approval of senior managers and practice partners. Such planning may be carried out at the beginning of each work stage or project phase.

II   Work supervisors have an operational view of planning

This is the most detailed and precise level of planning. The people who are directly responsible for design production get together to discuss their working methods, priorities and plan of action. This may be month-by-month, or stage-by-stage. The choice of people taking part in these discussions, however, is a tactical decision, which may depend on the scale of the project and the degree of involvement of each company, or department, in the project.

I   Design staff are concerned with task planning

Individuals or small groups who work together, in one company or department, agree their priorities and objectives, generally week-by-week. They also consider how each member of their team should co-ordinate their work with the others.

5.12   PROJECT OVERVIEW AND THE MASTER PLAN

A typical project master plan, as illustrated in Figure 4.1 (page 37), shows the sequence and duration of the main stages of a project. The more abstract view given in Figure 5.3, recognises that at the outset people will be thinking about the desired outcome and their immediate tasks, but they may not give much thought to the intermediate stages of detail design and construction. This is a reasonable way of looking at the coming programme of work, since the detail design and construction stages cannot be planned until the aims of the project have been clarified and an outline design has been produced.

As work progresses, the viewpoint of the planners, managers and designers also moves forward, so that the hills and valleys stand in different places, relative to their changed viewpoint. The way that this view changes has been referred to the 'rolling wave' of planning and execution.[7]

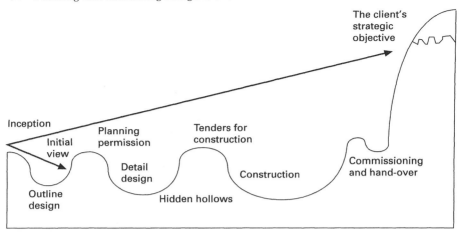

Fig. 5.3   Initial view of the project

At the beginning of each recognised phase of a project, its ultimate objective may have to be reasserted. The way through the next valley should be planned, before the design team leaders get too deeply involved in the work of that stage, when they may lose sight of the more distant project objectives.

This analogy also illustrates the tendency for each phase to begin with a confident rush down–hill, followed by a loss of momentum half way through. The subsequent 'climb' to complete each phase can be hard work.

Figure 5.3 makes the point that a building is rarely the ultimate objective of a construction project, as far as clients are concerned. They will be interested in looking beyond that stage to a more distant 'mountain': the use, letting or sale of their buildings to achieve a primary aim. To give focused service to clients, it is therefore essential that the full implications of the project objectives should be thought through, by the client and the key adviser (which may be the lead design consultant), and brought to bear on strategic decisions about the design and the organisation of design work. If this is not done, detailed work planning may be invalidated, when fundamental decisions have to be reconsidered at a late stage in the work.

This illustration also shows that, as the project team embarks on each new phase of work, their view can easily be directed into the hollow in front of them, where they seek to anticipate and to preempt difficulties. By taking this view, the team may easily lose sight of their client's business objectives. This is one good reason for having a management hierarchy, whereby individuals at the higher levels have continuing responsibility to keep long-term objectives and business strategies constantly in view.

Communication between different levels of management may be ineffective because their focus of attention differs.[8] It is therefore recommended that the managers who working on a construction project, but in different companies or departments, should be in regular communication with the other managers who are on their particular level. This helps to maintain a common view of priorities and progress, as measured by the criteria most relevant to their seniority, whether it be policy, strategy, tactics or production operations.

5.13 STARTING UP THE PROJECT ORGANISATION

The context of initial design activity, in the developing project organisation, is illustrated in Figure 5.4. This shows that initial attention focuses on issues of policy and strategy. The term 'parameters' suggests the presence of boundaries. In this context, the parameters include cost and time constraints, together with the key design acceptance criteria worked out during the project definition phase. The design briefs set out the parameters of acceptable design work.

Construction projects are generally initiated at the highest level of a client organisation and the very first statements about the purposes of a project and how it should proceed relate very closely to their policies, that is, the primary purposes of the client organisation. This level of management will define the key parameters of the project, such as the total budget, the time-scale of work, or the market for the product (the building). Thus, the left side of Figure 5.4 represents work to be done by a small number of directors and senior managers.

Policies for contacting stake-holders merge into strategic considerations, such as the judgement of risks that may be run if certain groups are stirred-up or by-passed. Other strategies would be discussed by senior managers, such as the advantages and disadvantages of employing a full design team at the early stages of design work.

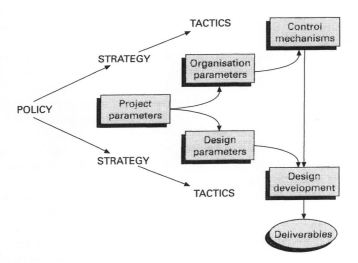

Fig. 5.4   The context of initial design activity

The procurement schedule for professional appointments should figure strongly in the start-up programme of work, which should allow sufficient time for short-listing and interviewing prospective practices, as well as negotiating terms of appointment. Designs are only as good as the designers and the advice they take. The early appointment of an adequate design team is indispensable to the success of a project. At minimum, design work for anything larger or more complex than domestic-scale buildings requires input from an architect, a structural engineer, a building services engineer and a cost consultant. Other consultants may be needed according to the nature of the project. It can be seen that, as the arrows in Figure 5.4 follow towards the right, work is being passed down through the managerial hierarchy. This also implies that more people become involved and that the project organisation grows, over a period of time.

For projects that are innovative, it is usually wise to employ a full set of advisers in order to avoid back-tracking on early design decisions, which might otherwise be found to be wrong at a later stage. On the other hand, if the proposed buildings are of a kind with which the client organisation, or their key advisers are very familiar, there may be no need for an extended design team in the project definition phase of work.

5.14   HIATUS BETWEEN SCHEME DESIGN AND DETAIL DESIGN WORK

Very often, there is a protracted delay between completion of the scheme design and the beginning of detail design production. This may be because of planning appeals or difficulties in securing finance. It may also be because the design is developed some time in advance of an anticipated market opportunity and the client is waiting for the best conditions to proceed. The instruction to commence detail design work may occur only when this much heavier commitment is reasonably likely to show the quick return that is necessary to maintain the financial liquidity of the client organisation.

It is often reasonable to regard the technology used to develop outline and scheme designs as totally separate from that needed for detail design work. This is a strategic viewpoint that reflects two factors. Since there is often a delay of months, if not years, in moving from the scheme design production to detail design work, information technology is likely to move on to something more sophisticated in this period. The later work of detail design often benefits from 'retooling' with up to date IT equipment. It is also recognised that, because the detail design production team is greater in number and complexity than that required for the work of project definition, communication has to be more sophisticated. This sophistication often extends to communication with contractors and suppliers, especially where construction is procured on a fast-track that overlaps the period allocated to design work.

One implication of the hiatus is that the team working on detail design production is often a completely different group of people from those who produced the project definition, the design brief and the scheme design. It is therefore very important that every document produced in the early stages of a project should be clearly written and cross-referenced, so that a new team can work from it, without making any serious misinterpretation.

5.15   CONTROL MECHANISMS

The basis of controlling the later stages of a project are laid down in its early stages, through:

- The various briefs.
- The emerging quality plan.
- The developing cost plan.
- The work breakdown structure.
- The project master programme and the procurement assumptions on which it is based.

The work breakdown structure is a key element in this, because it should take into account the development of the other control mechanisms. The programme of work follows from it and should include the activities involved in building up the project organisation and a database of information gathered from research, briefing, design work and the gradual expansion of the project organisation. Together with the quality plan, the programme describes how design work is to progress, including the framework of planned communication, between the members of the design team, the stake-holders and others.

All the control mechanisms listed above should be set-up so that, notwithstanding a possible delay before detail design work begins, continuity will be maintained. Some preliminary consideration should also be given to the additional control requirements of following phases, so that these are not compromised by early design decisions. The safety plan for construction is particularly important in this respect, because its requirements may impinge on fundamental design decisions. For example, if the concept design shows extensive areas of glazing, the outline design and cost plans (for construction and running the completed building) must take into account how this will be cleaned and maintained safely.

5.16   APPROPRIATE WORK BREAKDOWN STRUCTURE

As explained in Chapter 4, the work breakdown structure lists project phases, tasks, and who will do them, along with the key deliverables such as the tangible outputs of the work. In the hurry to show progress when design work starts, there is a tendency to avoid exploring the work breakdown structure in detail. Projects that progress without this formal approach incur the attendant risk that necessary tasks may be skimped or forgotten, because their significance in the progress of work has not been made explicit.

(RIBA) Project stage	A. Inception	B. Feasibility	C. Outline design	D. Scheme design
Process management				
Finance	Primary budget breakdown	Plan design phase cash-flow and controls	Refine predictions of project cash-flow	System to monitor value of work done
Project environment	Identify sites and stake-holders	Identify interests of stake-holders	Manage consultation with stake-holders	Obtain site and approvals
Project organisation	Define services of key adviser	Set up briefing information system	Quality plan for design production	Design project information systems
Contracts	Appoint key adviser	Define design team acceptance criteria	Pre-qualify detail design consultants	Appoint detail design team
Work programme	Establish range of completion dates	Analyse procurement options	Outline project master plan	Schedule detail design tasks
The building				
Cost	Analyse the project financial objectives	Define project success criteria	Parametric cost plan	Target elemental cost plan
Value	Ongoing (whole life) value management			
Building functions	Outline activities and space requirements	Research similar buildings	Collate activity data into brief	Detail design acceptance criteria
Technical factors	Provisional list of building systems	Research similar buildings	Collate technical factors into brief	Specify elemental performance
Design activity	Analyse alternative sites	List acceptance criteria for sites and concepts	Select and integrate design solutions	Refine design solutions
Design deliverables	Make preconceptions explicit	Generate alternative ideas and concepts	Outline design for client approval	Definitive scheme design

Fig. 5.5 Breakdown of project information from inception to scheme design

A work breakdown is complete only to a certain level of detail. Figure 5.5 shows a breakdown which is developed from the 'information packet matrix' devised for the Bond van Nederlandse Architekten (BNA)[9], which is equivalent to the RIBA in the UK and the Royal Incorporation of Architects in Scotland (RIAS). This Figure outlines the information needed to develop the project organisation (in the top half of the Figure), as well as the design activities (shown in the bottom half). Its focus is essentially on deliverables that can be recognised by their existence or state of development.

This kind of approach is essential for planning and monitoring. The information packet matrix formed the basis of the Dutch equivalent of the RIBA plan of work and has subsequently helped to structure their advice about formal quality management in architectural practice.[10]

The time-scale of work on each information packet is far from equal and does not start and finish as neatly as the boxes suggest. For planning and monitoring purposes, it is advisable to convert the work breakdown into a networked programme. This makes it possible to recognise high-risk nodes or pinch-points and arrange for administrative, design and technical staff to be available when specific tasks are reached in the programme. The time-scale of work is likely to remain uncertain because substantial rework may follow the progressive refinement of design requirements. Furthermore, opportunities to improve the value of the project are quite often discovered through conceptual design work.

5.16.1   Row headings in the information packet matrix

The original BNA process model divided organisation factors into finance, programme and organisation. In Figure 5.5, the organisation factors have been further divided, firstly to emphasise the dependence of a building project on its environment and secondly, to articulate the need for formal contracts between the project participants.

The row labelled 'Value' stands out because it is not sub-divided. This heading is not explicit in the BNA model. It is included to represent the need to check the feasibility of design ideas not only against cost limitations, but also to ensure that every element of the design will return value for expenditure. This comparison is often far from straightforward, for example, in determining how much is to be spent on windows, the relative value of several design objectives has to be balanced, including the needs to control temperature, glare, noise intrusion, condensation, natural ventilation, minimum maintenance cost, and the safety of maintenance. The comparison for the individual element also has to be balanced with the relative costs and value of other elements, for example, it would be inappropriate to buy expensive windows if this left too little in the budget to ensure that the walls and roof attain a comparable performance. Buildings, their content and their use perform as a system. The balancing of likely cost and required performance is a process that continues throughout the design process.

Design activity is differentiated from the design deliverables, to add depth to the plan of work and provide more monitoring points.

5.16.2   A further description of selected cells

In Figure 5.5, column 'A. Inception', is fairly self-explanatory, but quite dependent on the particular project. The key adviser may be assisted by other consultants, as mentioned in section 5.13, according to the difficulty of the project, in terms of the technical complexity, tightness of restraints, such as time and budget, political sensitivity and other special circumstances, for instance, the project might be overseas.

In the 'Technical factors' row, the building systems referred to can be many and various. These do not all have to be articulated at the outset, but often become clearer as the briefing progresses. Other similar buildings are investigated and, when some conceptual design work has been done, a list of systems might emerge. An initial list of required accommodation is developed into a 'bubble diagram' representing the size, shape and inter-relationships of activities to be accommodated.[11] This arrangement should respond to site characteristics, such as its outline, slope, micro-climate, the location of access points for both vehicles and mains services, views that can be exploited and so on. The pattern of major buildings services routes is then attached to the diagram with emphasis on the space-consuming provisions, such as lifts, stairways, air ducts and drain pipes. The enclosure of the building and the partitions between spaces form another system, which is usually 'squeezed' into rectilinear shapes.

Together with the walls, the pattern of activities accommodated by the building determines the loading on floors and the clear spans needed for floors and roofs. The structural system then takes shape to support these other systems. As design work progresses, more and more patterns emerge that have systemic characteristics, for example, in some areas silence might be needed, although noise is generated in others. Fire risks have to be handled as another system, often in association with security arrangements. Gymnasia, kitchens and smoking areas may generate air that should not be circulated to other areas. Humid areas are candidates for energy recycling. Later in the design process, finishes and lighting have to be considered as inter-related systems, together with the choice of materials to achieve a continuity of architectural style, plus signage, furnishing, etc. The definition of these systems is project specific and may vary according to the current practice of the various specialists who are engaged.

In column 'D. Scheme design', the word design in the cell 'design project information systems' should be taken both as a noun and a verb. It is necessary to design an integrated information system for the project, to set-up information systems for detail design work and to manage design information during tendering and construction. The structure and content of these systems is a major subject that is beyond the scope of this book, but it is necessary to appreciate that this is not the easiest of activity groups to get right and to complete. Experience of design work and communication within project organisations must be brought to bear, with considerable pro-active capability.

The last cell in the 'Design activity' row is the refinement of design solutions. This can be a very important preparation for subsequent work on the design of construction and assembly details. Study drawings are often produced at this stage, to show how the ideas represented in the outline design should actually be built. Calculation is usually necessary, to reconcile the dimensions of components and, in some disciplines, to prove that structures or buildings services elements, as shown in the outline design, can perform their intended functions. Work by specialists in designing for construction can begin quite early, if manpower is available to do this, and be developed through the selection and integration of partial or full design solutions which lead to the definitive scheme design shown at the end of the 'Design deliverables' row. The results of this work should also be registered in the cells of column D comprising, 'Detail design acceptance criteria' and 'Specify elemental performance'. Together, these outputs comprise the essential information required for the next phase of work.

Similarly, the final cell in the 'Work programme' row represents preparation for the detail design phase. Designers who have worked on the outline design can set down their understanding of the tasks that will be required to complete the detail design. This is useful, even if it is in a simple form, such as a list of required drawings and specifications. This information also relates to the top cell in column D, as a work breakdown should assist the value of completed detail design work to be estimated, as it proceeds.

5.16.3 Risk identification

Risk management is a specialised area, but nevertheless one that non–specialists cannot afford to ignore. If action is not taken to minimise the hazards associated with work, then disruption is more likely. Planning and monitoring are themselves a set of techniques that is applied in order to reduce risks, such as those that follow from disorganisation. The successful application of these techniques may depend to a large extent on the ability of the managers to recognise potentially disruptive events and circumstances and do something to reduce their likelihood and the possible damage they could do to the progress of work, its profitability and the quality of the finished building.

It has been said that the majority of significant business risks are associated with people, rather than technical or financial factors. This goes far beyond health and safety matters, to include dimensions of social psychology, such as how people interact at work. Among the many things can interfere with progress, poor communication is common, for example, where people make assumptions that reflect their point of view and fail to make messages clear and explicit for the recipient. The main concerns for the progress of design work may include inadequate information input to tasks and sections of work, a poor working method, unreliable designers, or inadequate checking of communications between them.

The theory of risk management makes much of quantitative methods, that is, to show which hazards are most likely to disturb the progress of work and which could do the most damage. However, the most practical aspect of risk management is to recognise where the risks are, in the first place. A subjective judgement of which risks are the most significant is generally sufficient to ensure that action is effectively prioritised. It is beyond the scope of this book to consider systematic approaches to the management of risk. It is only necessary, here, to point out that work planning and monitoring should feed information to the risk management system, if one has been set up, and that it would be used as an essential tool in operating such a system.

It is helpful if clients and project participants are all alert to the risks that attend their work. Factors they identify can be gathered together and assessed by the managers of the project (and each design company involved), to see the overall significance of each recognised hazard and to decide what to do about it. The information packet matrix offers help in doing this systematically for design activities. Separate studies may be needed to recognise business and other hazards that are outside the design process and therefore beyond the scope of this book.

Each column of cells in the matrix hints at a different set of potential hazards. Appropriate questions to ask may include:

■ In what way could the output of each cell be incomplete or contain errors?
■ What are the possible effects, further on in the project, of these omissions and errors?
■ Are there potential relationships between these hazards that could make them tend to knock-on from one to another, converting a minor hazard into a major one?
■ What is the potential for consequential damage to the quality of work, the programme, financial success and to the health and safety of the project team, future building users and anyone else?
■ What could be the most serious consequence?
■ What is the likelihood of this happening?

The introduction of the Construction and Design (Management) Regulations of 1996 made the health and safety aspect of risk management obligatory in Britain. The appointment of a planning supervisor is needed on all but the very smallest construction projects. This appointment should not be a substitute for attention to the other hazards to the success of projects, which may call for a different kind of expertise from the management team or a specialist adviser.

Risks can never be eliminated, but the most significant ones should be identified and prioritised for attention. There are never sufficient resources to address every conceivable hazard that might affect personnel, the progress of work and its profitability.

5.17   THE NEED FOR FLEXIBLE MANAGEMENT

The overall time allowed to get a building constructed, from project initiation to hand-over, often looks ample and so there is a tendency for people to let the initial stages run at their own pace, especially if communication with the client and stake-holders is slow. For this reason, it is important to develop the overall procurement strategy for the project and look carefully at the time that is going to be needed to obtain authorisations, finish the design and complete the building on site. A firm grasp is needed, both of the reasons for the client's construction programme and the difficulties of the design team, in order to push work ahead at the pace that is needed.

The time-scale of work, in the early stages of a project, depends heavily on communications with the client organisation, local authorities and others. A quick response is often needed from the design team, to rework briefing information and designs as and when the client or the authorities return their views on earlier work. It is rare for intense effort to be maintained for more than a few weeks at a time on any one project. Consequently, design consultants often switch rapidly from work on one project to work on another one, and then back again, so that their time is always used productively.

It is not unusual for some projects to be blessed with a generous timetable. An example of this can be found in government funded construction proposals, where the sums to be spent on a project over several years are determined in advance. These projects give opportunities to analyse the design and construction programme in depth and achieve outstanding value for money from the project.

5.18   OUTSTANDING INFORMATION – 'QUESTIONS AND ANSWERS'

Substantial volumes of information have to be handled, even during the early stages of construction projects. The extent and complexity of this information are often obscured by the appointment of experts who are very familiar with their specialised areas and therefore able to manipulate many factors in their heads, without apparent need to set it down on paper. Individuals in a good design team generally have worked closely with other experts in the past and understand substantial parts of each other's knowledge base. However, there is always a need for on-going clarification of the reasons why options are being considered or chosen and how other aspects of the design should develop in response.

The following list suggests some attributes of information that it can be useful to register, to ensure that there is a timely response to any need for information and that this is directed to the person or group who require it. Lists such as this are generally set out as standard forms, to ensure that individuals who make enquiries consider all relevant points.

- Each enquiry should be given a unique identification code, so that it can be established with certainty that responses refer to this enquiry and not some other enquiry.
- The general area of the enquiry, should be stated (such as, ventilation), so that it is easy to list all outstanding information in each area of work and thereby assess whether that work can proceed effectively, or not.
- The company initiating the query.

- The individual completing the form.
- The purpose to which the answer will be put, for instance, 'included in an elemental cost plan'. This enables the significance of the enquiry to the progress of work, possibly a whole chain of inter-related design and construction operations, to be recognised.
- Any known complicating factors, which could render the answer more or less certain and/or readily applicable. This may also hint at contingent areas of the design which might be affected by the answer, such as environmental conditions (for example, high humidity), maintenance and life-cycle cost implications.
- A location, or locations, if information is needed to design a specific part of the building.
- Relevant documents, which may include the brief, regulations, drawings, specifications, and calculations.
- The details of the information sought. This should be expressed as specifically as possible, for example: 'What is the type and duty of a specific item of plant?', 'What is the crushing strength required of blocks in a particular wall?', 'Precisely why does so-and-so believe that a separate room is needed for such-and-such a function?'
- The date when the information is required.
- The organisation/s to whom the question is, initially, to be directed.
- An individual's name, if the enquiry must be directed to them.
- Additional information that the person making the enquiry thinks may help the respondent to process it quickly and return a pertinent reply, for example: 'I saw this done in such-a-way in building X – would this be appropriate in this situation?'

For a question and answer sheet to work effectively, it should also contain:

- Suggested criteria for accepting the answer as fit for the purpose (if these can be predetermined), together with the authority of the person stating these criteria (the form may be initialled as an authorised enquiry, for instance, from a particular design specialism).
- Space for the names (or initials) of the people sending the answer.
- The answer, either on the same form, or attached and returned with the form.
- Check boxes to control the progress of the enquiry and the aptness of the answer. For instance: Query sent by; Date sent; Person answer is received by; Date answer received; Answer reviewed by (person); Date answer reviewed; Answer accepted as fit for its purpose by criteria 'A', 'B' and 'C'.
- A forward reference to any query or queries that were generated by the answer.

Question and answer sheets have been used for a long time in the construction phases of building projects, to co-ordinate the work of specialist designers with each other and with the administrators who order materials and organise work on site. This approach can be refined, to suit design work, and used from the outset of a project. This provides the additional benefit of an audit trail for design decisions. This mechanism can contribute to quality management and also as a reference for the people who develop the brief that is used to inform new participants who come to work on the project.

A question and answer system can highlight situations where design work may be proceeding whilst a significant question remains unanswered. This information can help to identify the risks of disruption to design work 'down-stream'.

This system should be integrated with correspondence and minutes of meetings, so that all answers and unanswered questions can be found in one place. This requires some centrality of management, or at least, an information node in the organisation. Information technology can readily be applied, for example, by networking the computers used by companies working on a project, with the benefit that information can easily be retrieved by everyone who is authorised to look at it.

During the design work, the project information base should be managed by the lead designers. As the project progresses towards construction, however, it may be appropriate to transfer the system to the leading construction contractor or manager. Alternatively, it can be run by a project support office throughout every stage of the project. The latter option has the advantage that particular individuals can develop in-depth understanding of the system and the information that it contains.

5.19   A NOTE ABOUT SMALL PROJECTS

Small projects can be as complicated as large ones, for instance in the detail of briefing information, technical factors to consider and the number of alternative sites. Every sub-division of the preliminary project stages, and the information packets referred to above, can be found in small projects and they are just as significant in this situation.

Small projects are often expected to be finished on a very short time-scale and it can be extremely difficult to approach them in a completely systematic, step-by-step way. While the technical information required for small projects may be as complicated and voluminous as it is for large ones, in general, the organisation and communication patterns are not. This means that managers of smaller projects can successfully use systematic approaches, such as those suggested in this chapter, in a less formal way.

SUMMARY

Whereas a project definition statement is usually written before design work begins, the design brief generally develops along with the work on design concepts and the outline design. Work on the detail of the design brief often continues even after this. Client personnel may sometimes be required to research their requirements in detail, in liaison with the design team. Such work may need skilful organisation. The briefing process continues throughout a project as new members of the design and construction team begin their work.

Early work on a building design is done by small teams. This work rarely needs detailed planning and monitoring. However, preparatory work on the organisation and control of design production and construction can begin at this stage. Cost control mechanisms are important from the outset. Quality control can be assisted by setting up a register of outstanding information. This should take the organisational design into account as well as design information, as shown in Figure 5.5 (page 78), the project information breakdown.

Change control and risk identification are essential to ensure that production plans are not invalidated or undermined by design changes or other unforeseen circumstances. These techniques are especially important where parts of a design are deliberately left indeterminate, as may happen with many building types that must be adapted to occupancy requirements that may be known only at the last minute.

The capability to control work is organisational. Clients who lack experience of building projects generally rely on professional advice. Relationships with project stake-holders may have substantial effects on the progress of work. Political, strategic, tactical and operational responsibilities must be carried at appropriate levels of the management hierarchy. Far-sighted leadership may be needed to ensure that the client's objectives are kept in view and respected by all the design team.

The engagement of specialist designers at the appropriate stage of the work may be critical to formulating and following a workable programme for design work, without back-tracking or other confusion. This generally means sooner, rather than later.

There is often a hiatus (a period when work is suspended) between the completion of the outline design (or the scheme design) and the commencement of work on the detail design. It follows that there is often a change in project personnel. It can therefore be important that all the early research work, design decisions and identified risk factors should be clearly recorded in the brief, or represented in the quality plan for the design and the project organisation, and communicated to the new management and design team.

NOTES AND REFERENCES

1. Gray, C.; Coles, E. J. 1988 The identification of information transfer between specialists to form a design process model – report on Science and Engineering Research Council (grant D/31034). University of Reading.
2. Vietsch, C. A. 1987 Anamnese, diagnose, therapie – een onderzoek naar de bouwvoorbereiding van algemene ziekenhuizen –PhD thesis. Eindhoven Technical University (Translation: Sickness, diagnosis and cure – research into the design development of general hospitals.).
3. Beheshti, M. R. (Ed.) 1985 *Proceedings of the International Design Participation Conference.* Eindhoven Technical University.
4. Gray, C.; Hughes, W.; Bennet, J. 1994 *The Successful Management of Design.* University of Reading.

5. Vietsch, C. A. *op. cit.*
6. Churchill, W. S. 1959 *The Second World War*. Cassell and Co. Ltd. First published 1948, abridged version completed in 1959.
7. Association of Project Managers 1988 *Proceedings of Internet World Congress of Project Management*. Glasgow.
8. Dodd, J. 'Organisation structure of the design team' in Gray, C.; Hughes, W.; Bennett, J. 1994 *The Successful Management of Design*. University of Reading.
9. Bond van Nederlandse Architekten (Translation: The Institute of Dutch Architects.) *Informatie-ververking Tijdens het Bouwproces*.
10. Bond van Nederlandse Architekten 1995 *Kwaliteitszorg voor Architekten*, 2nd edition.
11. Jones, J. C. 1981 *Design Methods – Seeds of Human Futures*. Wiley Interscience.

CHAPTER 6

INTEGRATED PLANNING OF DESIGN AND CONSTRUCTION

6.1 INTRODUCTION

Alternatives to the traditional system of construction procurement have developed, so that buildings can be designed and constructed in a shorter time. These require more consideration to be given to the sequence of the design work and the relationship of this work to construction operations.

This chapter focuses on the organisational framework in which detail design work is carried out. One important issue is how the design input of specialist suppliers and contractors is handled. Only the most commonly used approaches are examined, to illustrate the principles involved.

6.2 PROCUREMENT STRATEGY

The choice of procurement method depends on factors such as the relative importance to a client of control over the cost of building works, the quality of design, materials and workmanship and the time-scale of completion. Figure 6.1 shows four options in outline, to show how the various approaches to overlapping the design and construction phases of work permit completion within different time-scales.

When early completion is important, the overall time taken by the design and construction can be reduced by beginning the construction before the design work is finished. Many procurement methods have developed, from the traditional approach, where the design is virtually finished before construction work begins, to the 'fast-track' system that tries to achieve a maximum overlap of the design and construction phases. Five typical procurement systems are:

1. *The traditional approach*: This aims to ensure that design work is complete, before a bill of quantities is compiled and competitive tenders obtained from multi-trade contractors, one of which is selected to carry out all the construction work. This approach can give good control over the quality of work, provided that construction operations are well supervised. However, the whole process of detail design, followed by billing, tendering and the construction operations

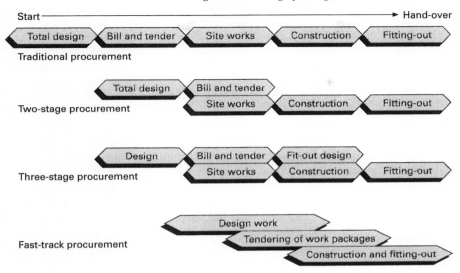

Fig. 6.1   Comparison of procurement time-scales

tends to be very long. Control over cost is dependent on the general economic environment, as in periods of high inflation construction prices can rise significantly during the detail design and tendering stages. This system may be unsuitable for commercial projects, especially where the decision to create a new facility is taken as late as possible in relation to a 'window of opportunity' in the client's operating market.

2. *Two-stage tendering*: In this system a contract for the preparation of a site and the construction of foundations can be tendered in advance of the main contract for the superstructure and fitting out. This enables work to start on site several weeks or even months earlier than would be possible with the traditional approach. During this time, the detailed designs for the building are completed, the bills of quantities are drawn up and the tendering procedures followed to establish the main construction and fitting-out contract.

3. *Three-stage tendering*: This is similar to the two-stage system, except that the detailed design of fitting-out design work and the related tendering process are put into effect after the main building work has started. This is particularly suitable for buildings that are not let until the construction phase is under way, which means that the identity and particular requirements of the tenants are not known until late in the programme.

4. *Fast-track procurement*: This approach divides construction operations into work packages that are the subject of separate contracts. To gain the potential benefit of a reduced overall duration of the design and construction period, the design information for each of these packages has to be completed to suit its specific tendering deadline. This approach requires very careful planning and control, to ensure that the design decisions made for contracts that are let first do not

compromise the scope for completing good designs on the later packages. In most cases, this approach is only followed when an organisation can provide specialised project managers, design managers and contract administrators who have experience of this way of working. Projects that are run in this way can be classified as employing a construction management or management contracting approach. These labels are subject to change and new ones can arise to describe variants and further development on the theme.

5. *Design and build*: This is not illustrated in Figure 6.1 because design work and construction in this procurement approach can be overlapped in several different ways. This approach is primarily distinguished from the others in that a construction company is engaged to complete the design as well as to construct the building. This engagement can run from quite an early stage of a project, starting with the detailed design (RIBA stage E), or even including the scheme design stage. This arrangement is more likely to be adopted where the building type is common and the design requirements are conventional. Normally, builders that have achieved previous successes with the particular building type would be engaged.

6.3   THE GENERAL PRECEDENCE OF DESIGN AND CONSTRUCTION

The manner in which detailed design information is developed and fed into the construction process is shown in outline in Figure 6.2. The capital letters on the left indicate an approximate alignment with the RIBA Plan of Work definition of project stages (see Chapter 3, page 26). Stages G and J are omitted to simplify the picture, as well as the design and construction of external works and landscaping.

Starting at the top left corner of Figure 6.2, the architectural arrangement of space determines the spans of the floors and the roofs and the main routes of the services installations. The precise details of the structure would then be developed from this. Note that detail design work develops, broadly, in the opposite order to that found in the construction operations.

On site, the structure is built first, followed by the services. The architectural elements are then constructed in two main phases. Wall and roof coverings first, which seal the building from the weather and allow the services to be installed. The partitions, fixtures and finishes are then fixed to the structure, clear of the services.

This general sequence limits the potential for overlapping detail design and construction, because some design decisions that naturally come late in the development of the detail design, are needed at an early stages in the construction. Problems with design co-ordination generally become more difficult to deal with where the time-scale of design and construction is reduced. This is discussed in more detail below.

The final details of many components can be dependent on actual site dimensions, which can limit the value of generating production information in advance. This restriction can be overcome by the careful management of

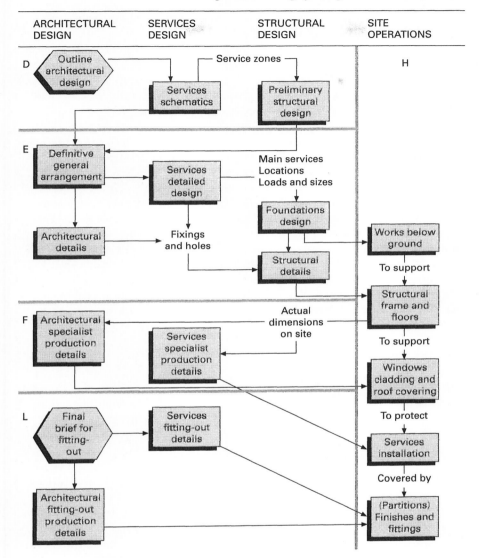

Fig. 6.2 General precedence of detail design work and construction

dimensional tolerances such as the 'looseness' of fit between components. Where possible, the component parts of a building are designed so that either they can be fixed together over a range of positions, for instance, by the use of dove-tail slots that allow bolts to be put in anywhere along their length, or so that the parts can be cut on site to the required size with fixings that are drilled or welded anywhere in a predetermined area. Nevertheless, the inter-relationship between the generation of the final production information and the implementation of the building construction is such that progress can be significantly impeded unless the programme is adequately planned in advance.

Just as tolerances must be allowed in the dimensions and fixing arrangements of components, the internal fitting-out has often to be considered a 'grey area' in the design until a late stage in a project. Figure 6.2 shows phase L to be disconnected from the earlier design work. However, this is a little deceptive. If scope is to be left for making late decisions, for example about the choice of finishes and the positions of partitions, a very clear design strategy is needed early, to allow this to take place within the building elements such as the structure, the envelope and the main services installations that will, by then, have been constructed.

Late fitting-out decisions are normal in buildings such as offices and shopping centres, where areas are let to tenants as construction proceeds. Late decisions are common in many other building types, to take advantage of market fluctuations and changes in technology during the period of the construction operations. For example, the final choices of information technology installations, security systems, medical equipment in hospitals, and baggage handling systems in airports may be left as late as possible to be able to incorporate the most recent developments. Even in housing, decisions about the fitting-out of kitchens and bathrooms are often delayed to the last minute to take account of the house buyer's individual requirements, which are frequently influenced by changes in fashion.

6.4   THE PLACE OF CONTRACTORS' DESIGN WORK

The relationship between the designers, specialist suppliers and sub-contractors can have substantial effects on the programme of design work. The following arrangements illustrate this point.

6.4.1   Traditional procurement using nominated sub-contractors

Thirty years ago, the technology of building construction presented fewer options. A competent professional design team, normally comprising the architect, a structural engineer and a services engineer, could complete most details of a design themselves. However, some specialist input, for example to detail lifts and stone facings, was often needed. The architect would nominate companies to provide the special items and services to the main contractor. These companies, being certain that they would be paid to carry out the work, could provide all the necessary design information in good time. Figure 6.3 shows the place of such specialist support in relation to the work stages, as defined in the RIBA Plan of Work.

Architects eventually became shy of the legal liabilities that attended this approach and learned to accept that specialised design input would sometimes be delayed until the main contractor had appointed the sub-contractors. The architect may insist, however, that each sub-contractor be chosen by the main contractor from a restricted list of companies of which they approve. The present forces of competition in the construction industry are such that most specialist suppliers are

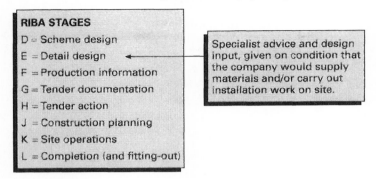

Fig. 6.3   The traditional place of design input by nominated specialists

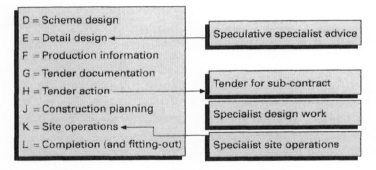

Fig. 6.4   Design input by specialist contractors, if not nominated

obliged to give some design advice to the design team in advance of securing a contract, to ensure that they will be included in the tender lists.

6.4.2   Traditional procurement with sub-contract design

The traditional approach is modified when the specialist suppliers and the sub-contractors are not nominated and the main contractor is given responsibility for all the sub-contractors. The pattern that this follows is shown in Figure 6.4 and it is immediately evident that the design work cannot be completed before the tender documentation is issued. Incidentally, this brings some uncertainty into the pricing of the job.

By avoiding nominations, a professional design team can pass responsibility for the elements that they do not design to the main contractor. The postponement of specialist design input, until after the sub-contracts have been awarded, can result in a need for the professional design team to rework parts of their design to accommodate the specialist design input. The time taken in this reworking may delay construction operations and it can also adversely affect the profitability of the work done by the professional design team.

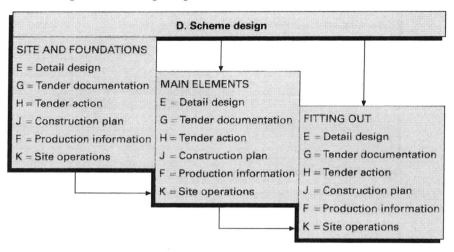

Fig. 6.5 Design stages staggered by three-stage tendering

6.4.3 Two-stage and three-stage procurement

Figure 6.5 illustrates diagrammatically how three-stage tendering permits design work to continue after tendering procedures and construction work have begun. Since two-stage and three-stage tendering allows the design and construction programmes to overlap, design decisions for some long-lead procurement items may have to be taken early in the design programme. It may be difficult to integrate the various aspects of the detailed designs soon enough to meet these deadlines, with the result that later modifications may be needed, either to the parts of the design that relate to these long-lead elements, or to the design of the elements after they have been ordered. Either circumstance can prove disruptive to the programme and the cost plan.

If it is intended that the contractor engaged for the site-works and foundations should also carry out the main contract, more elements of the design may be included in the tender at the first stage, to allow design work for sub-contracts to start in good time. The design information for such tender packages may be somewhat preliminary, which may also lead to subsequent amendment of the prices. This could make the work of the professional team more difficult, as they should try to avoid significant changes in the design after the documentation for the first tender has been sent out. Otherwise, it may become difficult to control construction costs.

6.4.4 Fast-track procurement

To maximise the overlap of design work and construction, a number of distinct deadlines are generated by a staggered programme for work package procurement,

D = Scheme design

J = Integrated planning of design, procurement and construction

E = Detail design

G = Tender documentation

H = Tender action

F = Production information

K = Site operations

L = Completion (and fitting-out)

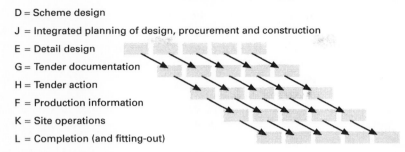

Fig. 6.6 Fast-track work proceeds in overlapping phases, rather than stages

as indicated in Figure 6.6. Some projects may generate over 50 such packages, although these would often be handled in batches. Fast-track procurement proceeds in overlapping phases, rather than the traditional approach of discrete stages.

This approach requires organisational arrangements such as construction management or management contracting to administer the complex contractual arrangements, plan the work flow and ensure that progress is maintained. It also requires exceptional design management skill, for a number of reasons. One of these is that, as designs are completed for work packages to be tendered, the critical design review for each package may have to proceed without full information about other incomplete design packages. Considerable experience may be required to recognise the potential implications of design decisions that have not yet been made. It can be important to reduce the 'spaghetti' of design information transfers by taking strategic design decisions that reverse certain flows, as discussed in section 4.18 (page 55).

As discussed in section 6.3 (page 90), the inherent difficulty in overlapping design and construction operations is that the design decisions tend to be made in the reverse order to those of the construction work. The structure is erected from the ground upwards and services are then fixed to it, whereas significant aspects of the design work develop in an opposite sequence, for example:

- The rising mains and the drains have to be located and sized before the basement and ground-floor structural design and water-proofing details can be finalised. To do this, the locations of, for example, toilets, plant, tanks, roofs, and major cable-ways, may have to be considered in some detail, although these services are installed after the structure is completed and weather-tight.
- Any holes for service runs must be taken into account when calculating the strength and size of all structural members that cross their path. Although the structure is built first, these elements must be designed first, or holes of adequate size left where they most likely to be needed.
- All fixings for lifts, plant, external cladding and heavy items to be connected after the structure is put up have to be taken into account when detailing the structural design. Items may have to be cast in place before the details of what must be attached to them are fully appreciated.

Consequently, there are likely to be difficulties with design co-ordination and detailed planning is needed to avoid problems on site. The administration can be simplified by certain design strategies that minimise dimensional conflict, for example, space zones can be reserved for the structural, services and finishes systems. The fixing methods and appropriate dimensional tolerances must always receive considerable and careful study.

The calendar time saved by overlapping design and construction may be at the cost of marginal increases in space and some over-specification of elements. This strategy can have a lasting 'spin-off' benefit, as clear zoning tends to make a building more flexible in use; for example, providing a generous space zone for the services may make it easier to reach pipes, ducts and cables for maintenance and to run additional services in future.

6.4.5   Design and build

The key advantage of procuring construction by a design and build contract is that close liaison with the building arm of the contractor's organisation can ensure that the design details and production information are appropriate to construction. This should assist construction operations to go ahead smoothly, quickly and with high standards of material selection, handling and workmanship. This often results in financial economies. This approach can be especially appropriate to system building, where many elements would be prefabricated in a factory to tried and tested patterns. In these circumstances, most of the detail design work would be done by the contractor's own personnel.

The client organisation, or their project manager, can make it a condition of the design and build contract that the professional team engaged in the initial stages of project definition and scheme design work should also be appointed by the contractor to carry through the detail design work. This arrangement, called novation, can ensure that the designers are employed who thoroughly understand the brief and will follow through with the design and quality acceptance criteria that have been developed with the client. Alternatively, a professional team can be selected and engaged by the contractor.

Where specialist design input is needed from suppliers, manufacturers or installers, these sub-contracts can be let earlier than in most other procurement systems. If this procedure is well organised, this allows their design input to be integrated into the overall design package sooner, as illustrated in Figure 6.7. Note that there is nothing to stop design and build contracts being fast-tracked, if the construction company has the required design management expertise, or can hire it.

The most frequently cited disadvantage of design and build is that early subjugation of the design team to convenience and economy in building has been known to result in designs that lack quality, particularly in their aesthetic appeal. But this need not be the case if the initial work of briefing and outline design is well done and the contractual arrangements ensure that these initial statements of quality are not compromised.

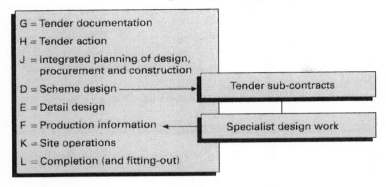

Fig. 6.7   Timely specialist design input to a design and build contract

This exemplifies the principle that the planning and monitoring of design work cannot deliver satisfactory results on its own, but should be integrated with an appropriate strategy in setting-up the project organisation. A full set of controls is needed, over the process of setting-up the project organisation as well as over the design work itself, as illustrated by the information packet breakdown shown in Figure 5.5 (page 78).

6.5   PROJECT INFORMATION DURING DETAIL DESIGN AND CONSTRUCTION

As explained in section 4.2 (page 36), the master plan provides a 'top level' breakdown of work that shows only the main phases of a project.

Figure 6.8 shows a 'second level' breakdown of work at the detail design and construction phases. This is set out in a similar way to Figure 5.5, which was discussed in section 5.16 (page 77). It is important to appreciate that the organisation of work, described by the top half of these figures, is important to the smooth production of design information, described in the bottom half.

The sequence of phases, represented by the columns in Figure 6.8, depends on the procurement method in use (as described in section 6.4). Note that the RIBA stages G (bills of quantities) and H (tender action) are present as rows rather than columns under the headings of 'Contracts' and 'Cost', because with many procurement methods these activities proceed simultaneously with construction planning and the production of design information for construction.

The breakdown given in Figure 6.8 should not be regarded as definitive or correct. Work on real projects can never be divided as neatly as this kind of figure would suggest. Such analyses are tools for work planning and not the answer to all the possible questions. The generation of boxes like these helps to recognise aspects of a situation that should be considered, but no real situation really comprises this number of aspects, nor can it be neatly squeezed into four columns and eleven rows. Care should be taken not to assume that a work breakdown is complete at any

(RIBA) Project stage		J. Plan construction	E. Detail design	F. Production information	K. Construction
Process management	Project environment	Detailed permissions. Marketing and finance restrictions on working methods on site		Liaison with utilities	Official monitoring and stake-holder liaison
	Organisation	Safety plans	Team induction	Brief site managers	Communication
	Work programme	Establish design deadlines	Schedule and monitor detailed production	Co-ordinate specialist input	Monitor progress and schedule commissioning
	Contracts	Tune main contract and sub-contracts	Tenders and negotiated agreements	Boundary definition of supply contracts	Facility management and maintenance
	Process costs	Plan management overhead cash flow	Design team internal resource controls	Producers and suppliers controls	Monitor and adjust site supervision
The building	Cost	Bill quantities and analyse activities	Further breakdown and change control	Design to budget and control changes	Monitor/record agree/determine
	Value	Operational quality plan	Preliminary design review	Critical design review	Monitor quality of work done
	Building quality	Define tests and acceptance criteria	Acceptance criteria for performance	Check designs and control interfaces	Commissioning and hand-over
	Technical factors	Construction method models	Element and system acceptance criteria	Adjust to site dimensions etc.	Acceptance tests
	Design activity	Integrate methods of construction	Design and co-ordinate details	Component and assembly designs and schedules plus packing and assembly instructions	Compile information As-built for safety and maintenance files
	Design deliverables	Detail testing methods	Drawings mock ups and specifications		

Fig. 6.8   Project information during detail design and construction

stage of a project, or level of detail, because, when planning, it is all too easy to overlook tasks that should be taken into account.

6.6 FURTHER DESCRIPTION OF COLUMNS AND ROWS

Most of the boxes in Figure 6.8 refer to activities that do not have design information as output. They are, nevertheless, significant to design production and the use of design information for construction.

The planning of the construction activity is shown as the first column because it should determine some aspects of the detail design work programme. In other industries, planning for production has a substantial effect on designs, as this can make the difference between meeting or missing production deadlines, quality acceptance criteria and production cost targets. Where building construction is the least bit innovative, this policy should be followed to obtain comparable advantages (See section 3.10).

The essential distinction between columns E and F is that detail design work is mainly done by professional consultants, whereas the bulk of production information is generated by the specialist suppliers and the sub-contractors who manufacture the building components or carry out the construction and installation work on site. It is possible for a professional design team to provide a full design service, detailing the building down to the last nut and bolt, but this very rarely happens because the contractors and specialist suppliers are more familiar with this level of detail and can produce information more economically and, in many cases, with a more cost-effective design solution.

To understand the main implications of Figure 6.8 for planning and monitoring purposes, it is not necessary to explain every box in detail but the following should give an appreciation of the figure:

1. *The organisation at stage E*: Team induction refers to the briefing of every section of the production information design team. Many members of this team will be new to the project since they are brought in as the staff of the companies who have won sub-contracts for the supply of component parts or the execution of the installation work. Some of the briefing is conveyed through tender documentation and pre-contract meetings, but a broader briefing, by presentations and workshops, should be given on larger projects and for key sub-contractors. These events can have a social aspect, to enable team members to get to know the people with whom they should liaise in other companies.
2. *Organisation at stage K*: Communication is of essential concern at every point of a construction project. It is mentioned here to highlight the special difficulties of co-ordinating the work of specialist designers engaged on sub-contract work.
3. *The 'Contracts' row*: These activities are aimed largely at ensuring every party to the project knows what is expected of them and that no task remains unallocated. Outline and detail designs need to be analysed to establish the scope of design work for production and the boundaries between the work to be done

by different companies or design teams, including the project's design consultants.

The term contract includes areas of activity that may not be mentioned in a written building contract, but which are part of the normal expectation of co-operation between companies and individuals. This area might cover working methods, for example, or agreed dates when design work should be passed from one company to another.

4. *The 'Process costs' row*: This refers to the resource controls that are dealt with in Chapter 8.

5. *The 'Building cost' row*: Building costs should be tracked and analysed interactively with design work. Every client wants tight control to be exercised over the project cost. This requires a system to be set up to maintain a running check on the cost implications of design work as it develops. Boxes in columns E and F, the detail design and production information phases of the work, mention change control specifically, but design is a change process, so control over cost should be an integral part of the control system at all stages. As with other areas that require control, this is done by comparing the current situation with a plan worked out in the early stages of the project.

    The last stage of cost control proceeds during the construction phase, by means of continuing checks on the value of work that has been carried out. Design information can play an important part in this reconciliation, where this is the basis of agreed prices.

6. *The 'Value' row*: This is very important to planning detail design work and the production of information for construction. This represents the system of quality control set up to ensure that what is built will conform to the original design brief and to the detailed elemental performance and acceptance criteria that developed from it. The formal processes of design review are shown at stages E and F. In practice, design reviews can be difficult to organise when different aspects of a design proceed at varying rates. The most obvious case is when fast-track procurement is employed and the need is for exceptional professionalism by everyone involved; designers, managers, cost controllers and often the client representatives as well.

7. *The 'Building quality' and 'Technical factors' rows*: These explore more deeply the mechanisms that ensure the actual construction and installations fulfil the intentions of brief and outline design.

8. *'Design deliverables' row*: Designers need to reserve some time to correct the design drawings and specifications to accord with what is actually built, since, invariably, differences occur in the detail of many areas of construction work. The as-built information is used subsequently in operating the building and when major repairs or alterations are carried out. This information would include the safety file, which becomes the property of the client.

Techniques for developing this work breakdown into a programme of design work are described in the next chapter.

SUMMARY

Design advice from specialist contractors is often needed early in the development of construction details, so that other aspects of the design can take this into account. It may also be necessary to select contractors for specialised work as soon as possible, so that they have time to develop their areas of the design, in liaison with the project design team, and organise complex manufacturing processes.

It is now rare for specialist contractors to be nominated by architects. Instead, specialist design advice can be obtained at an early stage of design development by tendering specialist work packages. This principle has been extended, in recent decades, through fast-track procurement of building construction, whereby packages of related work are tendered at times to suit the programme of operations on site. Deadlines for the production of design information follow from this procurement programme.

Despite the existence of such approaches to the procurement of construction, several factors may still limit the flow of design information:

- Clients may wish to decide certain aspects of the design as late as possible, in order accommodate changes in the technology of their business, or the requirements of potential tenants or purchasers of the building.
- The general sequence of construction operations is different from the sequence in which detailed designs are normally developed. This tends to limit the amount of time which can be saved by starting operations on site before building design information is complete.
- The dimensions of some components can be determined only when the actual dimensions of other elements are known, in particular, the structure 'as-built'.

It is possible to accommodate indeterminate aspects of the work, for example, by allowing additional space for building services and finishes. Dimensional tolerances should also be carefully calculated, or fixing methods can be chosen that do not require the precise locations of components to be known in advance.

The programme for developing co-ordinated design information has to take these factors into account. If a fast-track strategy is adopted, the process of design production should also be designed. This demands additional skills from the managers and that the design team should take the implications into account in the actual design, that is, in their ability to freeze decisions in a logical sequence, while accommodating uncertainty in areas of the design (and site dimensions) where necessary.

CHAPTER 7

DETAIL DESIGN AND DESIGN FOR PRODUCTION

7.1   INTRODUCTION

Particular problems are likely to arise from the fragmentation of the design team, whose members work for different companies and in different places. This chapter develops a systematic approach to integrating their work, through the steps described in Chapter 4: listing required deliverables, analysing the dependencies of design tasks and determining a rational sequence for the work, in relation to the construction programme. The processing of construction contracts that include design work also has to be integrated into this planning framework, including the approval mechanism.

This chapter explains how the status of designs for production changes, as they are developed from the consultants' detail designs and specifications into an integrated set of instructions for production and assembly. Formal control mechanisms rely to some extent on informal self-regulation by the designers. A system of question and answer processing can allow this to be monitored to some extent. The formal measurement and reporting of progress to senior managers is developed further in Chapter 8.

To co-ordinate the designs contributed by specialist contractors, preliminary information has to be passed between them and the design consultants. On large projects, the volume of design information produced by sub-contractors, for the manufacture and assembly of their products, is generally greater than that produced by the team of design consultants (the architect, structural engineers and services engineers). The design programme of specialist contractors is an area of considerable complexity that usually runs simultaneously with site operations. It is, therefore, essential to consider how the design work undertaken by sub-contractors fits with the overall pattern of work.

7.2   FRAGMENTATION OF THE DESIGN TEAM

The various professional and specialised designers contributing to a building project generally work for different companies and often on several projects at a time. Each company has its office at a different location and individual designers may not consider themselves to be part of the team for any particular project. The main concern of their managers is to smooth the flow of work within their company, by looking at the criticality of timings and the float allowed to them in the work programme of each project. This orientation tends to dilute the commitment of each specialised group to deliver their contribution on agreed dates. In turn, this can make it difficult for the managers of each project to optimise their design programmes and ensure that each contribution to this programme is delivered on time.

The result can be compared to a set of gears with mismatched or missing teeth. Forward progress can be pushed so far, but it slips when a significant part of the design work is not produced on time by one company and is not available in time as input to further design work in another company. If the dependent company is doing a lot of work for the particular project, their designers can often be kept busy on another aspect of the work. When the required input becomes available, the delayed design work is prioritised and the dependent output may still meet the programme.

Where staff cannot be kept busy on other aspects of the same project, they may be diverted to other work for a while. It may then be difficult to disengage this staff at a later date, to catch up with the work that was delayed. In terms of process control, the calibration of the design team may be lost, through stop–start attention to the job, and an effort has to be made to ensure that the designers are correctly refocused on the objectives of the particular project and their design commission.

General management theory suggests that the work of fragmented production units requires administrative structures and integrative mechanisms to co-ordinate their activities and keep this in line with management strategies.[1] Effective formal and informal communication is essential to integrate the work of the various designers and design teams. By bringing designers together at an early stage to plan how to co-ordinate their work, communication can be established on a purposeful basis. The key information transfers can be scheduled, and their content and form discussed. Uncertainties can be identified, together with any risks that are carried by some, if not all the contributors to the design work. With competent management, or simply if personalities are complementary, this can develop into effective teamwork.

Special skills are needed to achieve integration, not least, administrative expertise and a strong awareness of how people communicate and organisations function. Teamwork, however, requires something more, that is, a leader with the ability to impart a sense of common purpose and identity. It can be particularly difficult to lead effectively in a situation where designers are working in different places and for different companies, with their own professional and business interests that may be quite different from those of the project's client and stake-holder group.

7.3   SELF-REGULATION

The full scope of information required from each designer, or the design team, is not usually obvious, but has to be developed through discussion. This can be assisted by the use of question and answer forms, as described in section 5.18 (page 83). As the dependencies between various design tasks become apparent, they can be developed into a planning network and production schedule that can be discussed, refined and agreed between the leading designers. This communication may be more purposeful if the construction team is also represented.

Co-operation and effort are expected from trained professionals, which means that the main value of a design programme is not in achieving close control, but in the way it enables teams to get on with their work in a co-ordinated manner, with a minimum of intervention. This approach minimises the managerial overheads, including that which might be needed to update an over-detailed design programme.

If the leading designers of each discipline consider in advance what has to be done, by whom and in what sequence, the consequence would be:

- The designers would understand when and how others are dependent on their work.
- An agreement could be reached that commits everyone to a timetable.
- Any problems that threaten the programme are more likely to be recognised and dealt with by those immediately concerned and with a minimum of pressure from managers.

7.4   DESIGN FREEZE

There are specific dates when each item of design information is needed, so that tendering, component manufacture and construction work can proceed on time. The professional team's work is generally considered as final when the drawings and the other documentation they have produced are approved by the client (or his agent) and the lead designer (usually the architect). To be as certain as possible that design information is complete, accurate and co-ordinated, the final approval of information may be linked to a critical design review, that is, a formal review process which would involve all the key members of the design team and, possibly, an expert agency that is not directly contributing to the design. Approval may be given in stages, as parts of the design information are completed, but this can leave doubts about the degree of co-ordination with the design information that has not yet been completed.

In fast-track work, design freeze dates are chosen so that all the design information that is relevant to a particular type of construction work is available to be be packaged into the tender for that contract. That is, the design freeze dates are tied into the procurement programme.

The traditional procurement method is a special case, where all the design information should be finalised, or frozen, in a period of a few weeks, in order to release it for measurement and inclusion in the bills of quantities.

It is usually expedient to freeze particular design decisions as the design work proceeds, so that dependent areas of the work can continue on the basis of definite information. A prime example of this is the base dimensional grid, on which every element of the building is set out from an early stage of detail design work.

7.5 CONTRACTS FOR DESIGN AND SUPPLY/INSTALLATION

Co-ordinative mechanisms in the construction industry have developed over a long period of time. Formal contracts can draw undue attention to risks and penalties, which can make it harder for staff to concentrate on the job in hand and the particular requirements of the project client. It is therefore important that the processes of contractor selection and the terms of each contract should be appropriate and well administered. If this is not the case, then this factor might, conceivably, interfere with establishing a suitable programme and the timely delivery of design services.

The sequence of steps in letting a sub-contract that incorporates design services is shown in Figure 7.1. A sequence of activities, such as this, should be built into the detailed planning network for each specialist design contract, whether or not this is integral with delivery of a construction work package. The procurement of each design service must be included as an element of the network, rather than an unlinked bar chart, so that the time-scale of design information delivery is calculated in relation to the work of the other designers and the ordering deadlines which ensure the timely delivery of components to the construction operations.

The steps may vary according to which work package is being considered and its specific requirements for information. In practice, the implications of the boxes in Figure 7.1 should be considered in greater detail. For example, mock-ups, specifications and samples of colours and materials may be needed. A wide range of considerations might apply to the box 'Order supplies etc. manufacture and install', including tools, special production processes, materials packaging, transport, and people to do the work.

The programme for specialist design production can add considerable complexity to the construction programme. Experience, insight and continuing discussion are needed, not only to get the designs right, but also to ensure that this work proceeds in a sequence whereby design decisions in each area support one another.

To plan production design work, that is, information intended for use in ordering parts, cutting materials and assembling the building on site, the analysis of required design outputs would be on the basis of the detail design information,

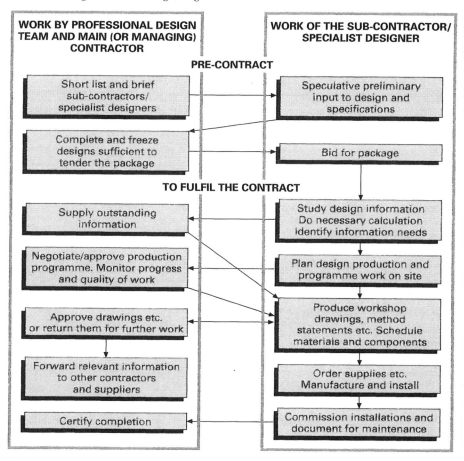

Fig. 7.1   Steps in managing and fulfilling a specialist design contract

supplied to the contractors by the professional design team. This might be supplied in two stages, as shown in Figure 7.1. The detail design would be regarded as frozen when tenders are invited from candidate contractors. However, it often happens that such information does not answer all the requirements of the contractors, so outstanding information has to be developed and forwarded to the contractors after they are appointed.

When a specialist designer wins a contract, the firm can draw attention to required information through correspondence and at formal and informal meetings. They must also carefully analyse the information that they are given, so that they can guarantee, as near as possible, that their work will not be held up by absent or inconsistent design or briefing information.

Where more than one contractor will be fitting items into adjacent, or even the same spaces, close co-ordination of dimensions, fixings, fire precautions, working methods, etc. is essential. It is not uncommon for unforeseen projections, such as flanges and valves, or the need for access to services installations, to be discovered only when items arrive on site to be installed. This often leads to last minute changes in the setting out of finishes, with all sorts of unforeseen effects on the appearance and utility of the spaces and equipment of the building.

7.6   LOOSE FIT AND CRITICAL TASKS

As described in Chapter 4, one aim of planning is to budget the resource requirements of work that has to be done, to ensure that the designers are available when they are needed. Another aim is to trace the dependency of activities, so that design work can proceed in a logical sequence. The critical path (the chain of dependent tasks that determines the shortest time-scale within which the work could be finished) is a focus of attention to rationalise the design and the sequence of work. The pattern of critical activities can vary significantly, according to the procurement method, as explained in Chapter 6.

Network analysis also shows the float enjoyed by the non–critical activities. Planning to overlap the design work with the construction operations can increase the overall availability of float. However, if building work has to be completed as soon as possible, the float is rapidly taken up as planners compress the programme. Their success in making a good programme can be measured by the emergence of a second, or even a third, parallel critical path. However, the greater the number of critical paths the greater the risk of a critical deadline not being met. If too much float is removed, this would also increase the psychological stress on people working on the project, to achieve the challenging targets set for time, expenditure and the quality of work.

Chapter 8 looks again at the use of float, to arrange work conveniently and give people a stream of work that improves motivation and concentration without increasing stress. As far as this chapter is concerned, the main significance of float is that, at least on the detailed levels of work analysis, the vast majority of design tasks do, in fact, enjoy some flexibility in starting and finishing dates. This means that they fit loosely into the work programme. This allows scope for design teams and individual designers to determine the sequence of their own work. This is a psychological 'plus point', since it reduces the likelihood of the stress that can arise if people feel that they are not permitted to direct their own activity.

The distribution of float across a work programme also has the practical advantage that, if work does not proceed entirely as planned, then the designers can use their experience (which is often very extensive) to adjust the sequence of their work to ensure that time is used productively. This self–determination is more likely to be facilitated within a programme that has been carefully thought about in advance and where managers, who are aware of the broad issues, tell their staff enough about work in the pipeline to enable them to make use of unallocated time and hours that are freed in an unplanned way.

7.7   LONG-LEAD PROCUREMENT ITEMS

Certain building materials and components must be ordered a long time in advance, so that they can be delivered to the building site when they are needed. If the construction industry is busy, demand for many items and materials can exceed the rate of their production and distribution. Where companies design the installations that they supply, a queue can also build up for the services of their design departments. These items include such elements as external wall cladding, lifts, air-conditioning ductwork and plant. Special finishing materials, such as purpose-made bricks and stone from a particular quarry, may only be supplied when notice of many months is given. Freeze dates must be carefully chosen for areas of the design that have to be co-ordinated with these special orders. It can be appropriate to consider such areas as *cooled* rather than completely *frozen*, so that limited changes might be allowed, but these should be made with considerable caution.

The sub-contractors or suppliers for long-lead items have to be chosen at an early date, but co-ordination problems can arise if they do their design work before other areas of the detail design are finished by the professional design team. These companies may be asked to accept some revisions to their designs at a later date, to take into account the production design work of other, related, contracts. The scope, however, of these revisions should be strictly limited, or tenders for these sub-contracts can be invalid and the logic of the design programme undermined. It follows that the scope of design freezes, relating to long-lead items, has to be very carefully considered at an early stage. The work needed to make all the necessary design decisions has to be identified and programmed well in advance, to ensure that enough time is available for the designs to be produced and that unwanted iterations do not occur.

7.8   ANALYSING THE DETAIL OF REQUIRED OUTPUTS

It is normal practice to list the likely requirement for detail design drawings early on in design work, to gauge the volume of work that must be done and the staff who will be needed, the level of skill and number of weeks required. This list is modified as work proceeds because the need for drawings and specifications will depend on the actual, integrated design, as it emerges. The need for production design information depends on what is to be constructed, rather than on the procurement method, but procurement will influence which company does what work and, sometimes, the sequence in which it is done.

The detailed list of required detail design drawings, specifications and calculations can be drawn up by searching all the available drawings for physical and systemic contacts between elements. This may be done using the output from the scheme design work stage. The analysis can follow a routine, for example:

■ Cross-sections make a good starting point because they are intended to pinpoint connections between elements. Each element can be followed in search of junctions such as: services to foundations; foundations to below-ground walls and damp-proof membranes; below-ground walls to ground floors; wall finishes to floors; wall finishes to walls; floor slabs to doors; walls to windows; walls to ceilings, and windows to ceilings.

■ The planner might imagine himself walking through the building using plans as a prompt. He could visualise each space, consider what it contains and how the parts will be put together, even going inside the elements of the building to find what the components are and where they meet, for example: door lock to door; door to frame; frame to partition; partition to light switch; conduits in partitions and ceilings; ventilation grilles to doors, walls, ceilings; connections between grilles and ducts; light fittings in ceilings, and suspension systems interlaced with ducts.

The requirements from each part of the design team, for example, for detail drawings, production specifications and parts lists, will be influenced by how responsibility is to be divided between the consultant designers and the designers employed by specialist contractors and suppliers. Each section of the design team should define the output required of them, systematically.

However the contracts for construction work are structured, detail design information is likely to be classified into sections that could correspond to tendering packages. This is convenient as a work breakdown, to clarify which consultants and specialists will contribute what design output, to recognise the dependencies between their work and as the basis of discussion about the programme of design work.[2]

7.9 LEVELS OF DETAIL

When analysing required outputs, it can be difficult to take into account elements that are not yet represented on the drawings. It may be necessary, therefore, to sketch-in levels of greater detail to be able to list the complete scope of required design work.

Design work generally progresses from large-scale to small-scale considerations. For example, room layouts at 1:20 can only be designed when their shape and size have been fitted into the overall plan of the building at 1:50 or 1:100. It is worth noting that the progressive iteration of design work, with ever increasing detail, corresponds very roughly with the traditional scales used for the drawings. This is because buildings comprise, and form part of, a hierarchy of elements, as shown below.

Scale	Building hierarchy
1:1250	The locality
1:200	The building plot and its environs
1:100	The general layout of the building
1:50	A suite of rooms or the services schematics
1:20	Room layouts, content and finishes
1:5	Detailed assembly and components

In general, when each successive stage of design work begins, the next level of detail exists only in the form of study sketches. These sketches (which may include calculations) can be usefully exchanged between the design disciplines, as preliminary information about where their work section interfaces are likely to be. This facilitates discussion about how they will be handled.

This generalisation that work proceeds towards levels of greater detail, can be reversed by the way in which CAD is applied. For example, in architectural design, the details of wall construction and the dimensions of doors and other components may be read from a library of detailed design information held in the computer memory. Items can be taken from the library at an early stage of design conceptualisation, and pasted-in to generate a detailed three-dimensional model of the design. This model can be complete with lighting, textures, furniture and people. This enables clients and other stake-holders to appreciate the design in a computer visualisation or a computer generated 'walk-through'. It also allows co-ordinating dimensions to be calculated readily, so that the designs for production may progress with more certainty. This is an instance where technology is revolutionising the way in which people work.

7.10   DESIGN PRECEDENCE MATRIX

Planning matrices are familiar to architects as tools to arrange the spaces in a building layout. This technique can also be used to analyse the necessary sequence of design work. It encourages the design work planner to review all the relationships between activities and reduces the possibility that important dependencies are overlooked.

Part of a design decision precedence matrix is shown in Figure 7.2. Each box, or cell, represents a potential relationship, where the work section, listed across from the row heading, is dependent on design decisions made for the section, as shown by the column heading. This technique can be applied flexibly and creatively, for example:

- It can be useful to differentiate intractable dependencies from those that could be worked around. The relative importance of the relationships can be differentiated by use of bold type, brackets, italics, colour coding etc.
- The detail of the analysis should respond to what the planners, on a particular project, believe to be necessary to tease out the potential problems that could occur in the flow of design decisions and information transfers between the design specialisms.
- Conventional network planning software can be used to assist this analysis, by entering data about the relationship of design tasks and decisions through the task precedence table, or directly as a Gantt chart if this display is preferred. Discipline is needed when using such tools, to ensure that potential relationships are checked systematically, for the answers are not in the planning programme, but in the current set of design information, interpreted with experience, imagination and focused interest.

Work sections	Ground-works	Concrete	Masonry	Steel frame	Curtain wall	Roof finish	Partitions	Ceiling	Joinery
Ground-works	ooo	**Foundation locations and sizes**	**Strip foundation locations and sizes**	**Pad foundation locations and sizes**					
Concrete		ooo	(Dimensional co-ordination)	(Dimensional co-ordination)	Insulation continuity and fixing	Upstands and holes for outlets			Precise opening sizes
Masonry	Damp proof courses and membranes	ooo		**Build out to conceal columns**	Interface details				Opening locations and sizes
Steel frame		**Floor loading pattern**		ooo	Fixing details				
Curtain wall					ooo	Parapet detail and insulation	Mullion detail	*Window head detail*	
Roof finish						ooo			
Partitions							ooo	Fixings and fire/acoustic seal	Opening locations and sizes
Ceiling					*Window head detail*		(Precise locations)	ooo	
Joinery									ooo

Note: Cells show that work sections named at the row headings depend on the sections named at the column headings.

Fig. 7.2 Part of the decision precedence matrix for design details (a multi-storey steel framed building)

If construction work is tendered in, say, fifteen sub-contracts, there would be 210 relationships to check (15 × 15 less the diagonal of the matrix, which is always empty). The dependency data could be input into a proprietary computerised network program in a couple of hours. However, it can take very much longer to sift through all the design information to recognise the significant relationships. It may be necessary to sub-divide packages and to analyse relationships in distinct areas of the building, as these are likely to generate different dependencies.

Since the design information for the contract packages is prepared by several design specialisms, the potential relationships between their work should be discussed between the leading designers in each discipline. The building services engineers can have a key role to play in this, since many of the elements they design will be sandwiched between the structural and architectural elements. Attention should be paid to the need for fixings, clearances, compatible appearance of exposed items, etc. Imagination is often needed to recognise relationships that are not between physical elements of the building. These might include attention to the practicalities of construction, as well as factors mentioned in Chapter 3 (section 3.5 page 27). Such analyses could easily absorb a couple of days of the leading designers' time. On larger projects, preparatory work may be done by specialist (work) planners.

The precise nature of the dependencies between design tasks should be recorded. Succinct notes, such as those shown in the cells of Figure 7.2, should be carefully drafted to indicate precisely the content of required decisions and how these should be communicated. Specific drawings or specifications might be needed to deliver this information. In such cases, the task bars on the computer generated Gantt chart can be coded with the document references.

Many dependencies will correspond to documents that are already scheduled for production, but supplementary sketches, specifications and calculations are often needed to explain the rationale of decisions to the other designers. Where the need for these can be predicted, they should be indicated on the network and the database of outstanding information.

The analysis should never be assumed to be complete; the entire design team should remain alert for dependencies that may not have been recognised. The register of outstanding design information, including question and answer sheets, should be monitored by a leading designer, to check for gaps and discrepancies that could disrupt or otherwise delay the smooth flow of work.

7.11   DESIGN PRECEDENCE NETWORK

When the activity precedence network that follows from Figure 7.2 is sketched, it becomes apparent that there is a tangle of mutual dependencies and loops, where one decision appears to depend on itself. This is a normal characteristic of design task planning networks and it implies that some of the work must proceed iteratively. It is up to those who are planning the work to impose order in such

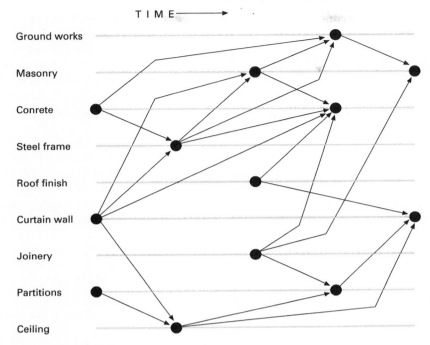

Fig. 7.3   Task network derived from decision precedences shown in Figure 7.2

situations, according to their judgement of what is the most practical way to proceed.

Figure 7.3 shows how looping dependencies can be clarified by separating out the design decisions that are needed first. Four cases of mutual dependency between work sections and two actual loops are unravelled here, by tackling four of the design tasks in two steps.

Figure 7.3 shows only architectural and structural elements. A complete matrix would include the building services elements as well. One common example of the interdependence between the work of the architect, structural engineer and services engineer is the location of water tanks. For example, a tank may be shown in the roof space on the outline design, but subsequent calculations indicate that the tanks would be too heavy for the roof trusses to carry. The water storage must therefore be divided to spread the load, or located elsewhere. At the outset, it may not be clear whether the structural design should be altered to carry the weight of the tanks, or if it would be better to spread out the water storage to suit the structure.

There may be many different locations in the proposed building where dependencies between the design of two particular packages exist. For instance, the structural engineer would need to know the sizes of all the air ducts that penetrate the floors and where they are to be placed, so that the designs for adjacent areas of floor can be sufficiently strong. When searching drawings and specifications for design dependencies, indirect relationships must be identified as explained above (for example, in section 3.5, page 27).

The full network of dependencies between design decisions and deliverables should be related to the procurement and construction operations. Careful analysis may help in deciding which decisions may be postponed until the specialist contracts have been awarded and may influence the sequence for tendering the construction packages.

Awareness of the matrix approach may help designers to appreciate that a very large number of dependencies can exist. More detailed analysis of design decision dependencies may be advisable, but this can, usually, be done informally as work proceeds. Team leaders should be pro-active in exploring links in good time, so that work priorities need not be revised when awkward dependencies are discovered.

7.12 PRELIMINARY, WORKING AND FINAL INFORMATION

Because many elements of a building cannot be designed separately, a work programme must show the key exchanges of partially complete designs. Important implications are that:

- Each professional discipline on the design team must be prepared to work from information supplied by the others that is not necessarily firm and final.
- A lot of design work will proceed on the basis of preliminary designs, supplied by members of other disciplines, which may be changed later.

Work planning proceeds on the assumption that work at the interfaces between designs (the areas where the work of more than one discipline must fit together) may be iterated twice during the outline design and twice again during the detail design process. This can be represented by attributing a different status to the design output as it completes each iteration.

The changing status corresponds roughly to work stages, as identified in the RIBA plan of work:

Work stage B: Preliminary design concepts are developed and evaluated, mainly to check the feasibility of the project.
This is followed by the first iteration:
Work stage C: Outline design proposals that are developed from preferred conceptual solutions in each of the main design disciplines. Although this can be sufficient to obtain town planning approval and usually enough to satisfy the project's financial sponsors, the design is not worked out in any detail.
Then a second iteration:
Work stage D: A scheme design is developed through inter-disciplinary co-operation, sufficient to obtain further statutory approvals for construction work to begin.

During the detail design:

Work stage E: The various designers develop further details of their designs on the basis of the scheme design, but the details of these partial solutions are not yet co-ordinated.

The first iterations of the detail design then occur:

Work stage F: Production information may be produced by the professional design team or specialised contractors. Some designs produced at this stage may offer an adequate basis for construction and manufacture to begin, but the information in different work packages tends to lack co-ordination.

Work stage K: This stage in the plan of work refers to professional design services during construction. Most design work at this stage is done by contractors and suppliers, in the form of detailed schedules, shop drawings and instructions for assembly, testing and commissioning. This is the final stage in the production of design information. As explained in section 7.19, however, further iterations may be needed to ensure complete co-ordination of this information, and compliance with requirements.

7.13   THE STATUS OF DESIGN INFORMATION

The status of all design information should be clearly marked, to ensure that:

- No design develops too far on the basis of preliminary ideas from another discipline, if these might change.
- Designers do not await more definite design input from another discipline, when they already have adequate information to proceed with their work.

The document (or computer file) management system should register preliminary and working transfers, in addition to authorised designs, so that the design managers in each discipline are aware of all the available information. This means that, in all but the simplest projects, even crude sketches should be marked with a date and a reference number and registered. The status of design information that is exchanged should be recorded, preferably the day when it is posted or transferred and on a schedule (or computer database) that is distributed (or accessible) to all parties, to ensure that everyone is working on the same, up-to-date, information.

It is traditional practice to give drawings revision numbers, together with the date, if they are changed and reissued. The status of different views, however, on one drawing may be different, as may the various parts of schedules and specifications. One way to clarify the status of views is to enclose areas which have not reached the general status of each drawing in 'clouds'. In the past, when designs were drawn on tracing paper, these clouds could be pencilled in on the back of the sheet, so that they could easily be erased, once the detail had been approved. If CAD and electronic data exchange are in use, then the status of different areas of the design, or layers, has to be marked in some other way. One of these is to print a table with each view, to show which parts of the drawing (usually separated onto different layers) currently have what status. The table itself may be held in several layers, so that this comment is unfailingly printed or transferred with the layer or view to which it refers.

Where specifications are word processed, different fonts can be used for sections having different degrees of finality, such as italic for preliminary, underlined for working and normal for final-approved. The increasing use of colour printers adds scope for differentiating the status of mixed information, as does electronic data exchange, where recipients of information will view it in full colour on visual display units.

These matters are essentially in the province of practical quality control and hence beyond the scope of this book. Nonetheless, it must be stressed that the implementation of quality control should be planned and monitored in parallel with the actual design work, because this is of major importance in ensuring that planning is practical and monitoring is meaningful.

7.14   CHANGING THE STATUS OF DESIGNS

The status of design information must be based on comparison with predetermined measures of acceptability. Professionals are considered to be experts, partly because they can apply their knowledge without having to refer to written authorities. In design work, this means that the majority of tests of acceptability proceed informally. Nevertheless, systematic attention to quality can benefit from the use of formal tick sheets to ensure that the criteria by which design information attains higher status are agreed and properly applied (see Chapter 2, Section 2.7, page 15).

Tests for the acceptability of design work should relate directly to the brief and the aims of the project. In effect, whenever one looks at a planning network, every link is, in reality, 'ghosted' by quality criteria that permit the products of work to be used in the next stage of the design or construction. In effect, this means that the application of systematic quality control must be planned together with the sequence of design work. For the quality management system to function properly, the necessary quality assurance information, such as acceptance criteria, should be developed in parallel with the network of design production operations. Since the non-availability of acceptance criteria can hold up design decisions and the release of design information, at any status, the professional designers should keep the development of this information firmly in mind when they are deciding the sequence and durations of design tasks.

To verify quality, preliminary work must be checked against the requirements of the brief, the intentions of the outline design and the current work of other disciplines. The completeness, co-ordination and legibility of working designs must also be thoroughly checked before information is approved as contract documentation. This may at times call for a third-party audit, carried out by qualified individuals, or a team, who are neither producing the designs nor involved in the building work themselves.

To achieve higher status, complete design drawings and documentation, or clearly defined parts of these papers, must be checked and authorised by someone who has been formally nominated as responsible. At certain stages, checks are made by members of different design specialisms working as a group, usually cost consultants, construction managers, the design disciplines and possibly someone from the client organisation. A preliminary design review of this kind is advisable during the detail design stage and a critical design review should be convened to sign off information that will be committed to contracts for construction.

If preliminary designs are given working status too soon, or working designs are considered final, errors will creep into the designs produced by others who have based their work on them. It may not be easy to trace all these consequential errors and the task of correction can be annoying and time-consuming. Some errors may never be teased out. Similarly, inattention to the status of designs can create unco-ordinated design output, if designers take preliminary work to be final. If the converse occurs, and final information is taken to be preliminary, time may be wasted if designers continue to explore design options that have, in fact, already been closed.

When contracts are let for packages that require further design input, for instance, by a sub-contractor, their shop drawings and information for ordering and assembly must be checked and approved for construction. The dates for this will relate to the time-scale of ordering, manufacture and delivery to site.

7.15 CHANGE CONTROL

Design work is a transformation process, so change is an intrinsic part of the work. In addition to design development that gradually fits the parts of a building together, changes can arise, through:

- Changes in legislation, or if unexpected interpretations are made of it.
- Variations in the availability of materials and components.
- Changes in requirements, such as revisions to a detailed level of the brief.

When any aspect of a design is changed, there are, invariably, knock-on effects to other aspects and it can be difficult to predict and limit the scope of corrective work to which this gives rise.

Some of the likely consequences of any change to designs would be:

- Delay in the completion of design information and possibly in work on site.
- Loss of co-ordination in design information.
- Reductions in quality due to haste.
- Altered construction costs.
- Erosion of design practice profits.

Some of these consequences can be avoided if the areas or elements liable to change are identified in advance. Efforts should be made to design both the building and the design programme so that changes to these areas or elements can be accommodated as late as possible. For instance, if no-one knows where office partitions will be required, the designers will need to ensure that window mullions, ceiling lights and ventilation grilles are arranged to accept them on a close module, and rather than planning to take partitions to the structural ceiling to limit noise transmission, a ceiling that blocks noise could be chosen. Many such problems have stock solutions which can be employed if the need for them is recognised early enough.

The cost implications of design developments must be monitored on a day-to-day basis. Changes of status from preliminary to working to final should only be made if the construction cost implications of the design have been verified and prove acceptable to the project cost plan. Life-cycle costs may also be considered, if a change to the design might alter the energy demands of the building, decrease the durability of some elements or amend the requirements for maintenance and cleaning.

In order to price developing designs accurately, it is useful to write detailed specifications as soon as possible. This differs from traditional practice, in which specifications were written on the basis of completed designs. Where CAD is in use, specification references may be added as designs are developed. This complements the measurement of quantities, which much CAD software enables to be called-off at any time.

It is important that the scope of permitted change is progressively narrowed as the design advances towards its final solution, ready to be approved for construction. The quality control system should keep designers aware of what scope remains for their decisions.

7.16   DESIGN CO-ORDINATION MEETINGS

Regular meetings are held between the leaders of each profession working on various aspects of the design. A typical agenda would centre around the design problems, but should also give attention to who will produce what output for whom, in what form and when.

The frequency and duration of meetings will be influenced by the complexity of the design and the thoroughness of previous planning of time utilisation and the management of quality. An awareness of the interdependency of work to be done by different parties can be generated at early meetings by giving sufficient consideration to the work breakdown and the key transfers of information.

As the design work proceeds, opportunities may be sought to reduce the dependency of one design decision on another. If they are inseparable, forethought has to be given to what information is needed to make the more difficult decisions, so that these can be made as scheduled. It is often advisable to register such information requirements on the list, or database. When unresolved questions must be passed on to a contractor, they should be contained in an explicit list of outstanding information, as this will help the contractors' designers to appreciate the scope of work that they are asked to do. Such outstanding information will normally have contractual significance, so this is another reason to define it as accurately as possible.

7.17   MEASURING AND REPORTING PROGRESS

Progress is measured, mainly, against the time and money remaining with which to complete a design project. Quality standards must also be taken into account, so that the real value of progress is maintained and work that has been registered as done does not have to be repeated.

This book mainly focuses on the programme of work. In this domain, progress is measured by comparing the work that has actually been done with the work that was previously scheduled to have been done by a given date. The methods of measuring and reporting progress are described in detail in Chapter 8, with the methods used to establish the value of work done. This information is necessary to ensure that sufficient money remains to pay the staff and overheads, as necessary to complete the work. In some instances, clients might link fee payments to measured progress.

Progress should be measured frequently, to assess the need for corrective action, as a shortfall can quickly escalate into an unrecoverable situation.

On a typical project, there are three levels of reporting:

1. *Designers* in each discipline report progress to their team leader.
2. *The leaders of each design team* report progress to the leading designer or the project manager.
3. *Senior staff* prepare summary reports for the client representative and persons to whom he or she reports, such as the financiers or a board of directors.

Progress reporting is not primarily concerned with what has been done. The focus is on the future because, rather obviously, the only remaining scope for corrective action lies in the future. Typical questions to be answered are:

- Is any individual, team or company struggling to deliver design work to the agreed schedule?
- Is any party going to be held up by late output from another, taking into account information awaited from authorities, the client, manufacturers and others?
- How can someone's work that is behind schedule be brought back onto programme?

Lists of outstanding information can be essential to the full appreciation of the actual progress of work, in comparison with the plan. To make it possible to identify relevant information, records in the database should include the dates when the information is required.

7.18   PROGRAMME CHANGES

Programme changes can be necessitated either by difficulties encountered within the project team or in response to external factors, such as a change in the availability of finance. In principle, they should be avoided. If companies have moved personnel from one project to another, a change in the programme of work can be difficult to overcome. A slow down in work is likely to leave staff under-occupied and a company may have to choose between carrying under-occupied staff, moving them to other projects, or terminating their employment. On the other hand, if work must be accelerated, sufficient additional skilled staff may be difficult to find.

If programmes must be rescheduled, everyone who could be affected by the changes should be consulted. Organisational arrangements for the generation and maintenance of work planning and monitoring are dealt with further in Chapter 9. Revised programmes must be fully explained to any of the designers within each profession who were not directly involved in the rescheduling.

7.19   APPROVAL OF SPECIALIST CONTRACTORS' DESIGN INFORMATION

Design work by specialist contractors usually requires approval from the professional team. It may also be necessary to seek approval from the local authority, for example, when structures have to be calculated or where there are implications for safety on site. The checking processes take time and there is always the possibility that errors will come to light, or a lack of co-ordination with other designs will be exposed, and time will be needed to put the necessary corrections into effect. Furthermore, clients or building occupiers are often consulted about internal finishes and fitments and they may wish to make changes. The sub-contractor's design programme must, therefore, be carefully planned and agreed with the project's leading consultants, so that it is properly co-ordinated with other work.

Design information can sometimes be prepared floor-by-floor or area-by-area, if this suits the construction programme. This can be especially relevant where working areas become available in a planned sequence. For example, external cladding and ventilation ductwork may be following on from structural work; roof finishes would follow roof installations; raised floors, suspended ceilings and wall linings would follow service installations. This can further complicate the management of work packages that have design requirements, as design work should be planned in detail to feed the production line of components. A construction manager would usually supervise the liaison of package contractors where design and construction are integrated at this level. However, contact is maintained with the professional design team that produced the detail design, for such reasons as the need to authorise the surface appearance and the detailed arrangement of the parts.

Figure 7.4 shows how information flows in one possible arrangement for controlling specialist contractors' design work. This cycle begins with design and specification work by the specialist contractor ('generate drawings, specifications . . .'). This information may take a lengthy path, via the main or managing contractor and the lead designer, on its way to consideration by other members of the design team and possible review by statutory authorities or the client. This may result in a change of status for the design that falls short of full approval. This diagram shows that specialist contractor's design information would

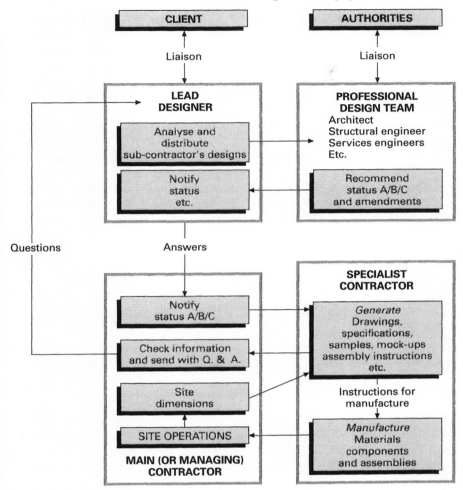

Fig. 7.4 Approval mechanisms for designs by specialist contractors

be reviewed in a minimum of six steps. Even if each cycle resulted in an upgrade in the status of the designs, the whole process could easily take 18 steps. If these each took one day, the design would be taking 18 working days to reach full approval. It is hardly surprising that, in practice, design information often takes months to be fully co-ordinated and approved for manufacture.

A secondary loop in Figure 7.4 links with dimensions taken on site, because these are likely to comprise an important input to a specialist contractor's design work. These measurements can be vital input to ensure that shop drawings will give accurate information for the manufacture of many components and assemblies, if these are to fit together precisely when they arrive on site. For example, the positions of fixings embedded in a concrete structure may have to be checked where components, such as external cladding, must fit together precisely, or where fixings are provided for internal finishing systems such as panelling, moulded ceilings or patterned tiling, or for the balustrade to a staircase. Although steel structures are usually erected with some precision, discrepancies from the designed dimensions of up to 100 mm have been recorded and concrete laid on site is even more prone to inaccuracy.

A common result is that final adjustments to designs cannot be made until parts of the building have been built. This starts the 'clock' in gaining final approval to proceed to manufacture and, in turn, this can be a significant 'brake' that slows construction work. In practice, it is often essential to minimise the potential problems and delays, by designing as many items as possible so that they can be fitted together, even if the conditions on site are not exactly the same as those given on the design drawings. This is a design skill based on an appreciation of what actually happens on building sites.

7.20   CONTROL OF SPECIALIST CONTRACTORS' DESIGN DEVELOPMENT

Approval to production information from specialist contractors can be given in stages, for example:

Status C: Designs that are unacceptable and require further work.
Status B: Designs that are generally acceptable, but where authority to proceed with the production of clearly marked areas is withheld.
Status A: Designs that are approved for construction in their entirety.

Care must be taken in awarding status B. There should be no lack of co-ordination that might require changes to any element of the design after it is approved for construction. In this event, the sub-contractor would have a right to claim additional costs or delay to the completion of works if the professional team alter any documented instruction.

The information flow, shown in Figure 7.4, represents a situation where the work of specialist contractors is controlled by a main or managing contractor. Their design output should be routed through this company, so that the construction managers are aware of its state of development. The information is then sent on to the professional design team for their comment and approval. The main or managing contractor may attach comments and questions to this information. The information is routed via the lead designer, so that this person or team can maintain a complete overview of design output and be in a position to comment on its co-ordination, when necessary.

If design information from sub-contractors travels in succession from one professional design team's offices to another, approvals and comments are likely to emerge weeks later, rather than days. Where samples and mock-ups are required, the design team may have to visit the site or the workshops of the sub-contractor. This has its own time implications.

A second consequence of serial approval, if this occurs, is that adjustments proposed by the different design disciplines occur at different times. Small differences can then creep into the design information being prepared in the various offices. Although most inconsistencies generated this way tend to be insignificant, if differing information begins to circulate in the offices of various designers and suppliers, the scope to generate confusion can be surprisingly great and costly to disentangle.

Design work to be submitted for comment or approval should, therefore, be sent simultaneously to the professional team, the main contractor, or managing contractor, and the other sub-contractors working in the same area. This is an important justification for project managers to insist that all offices operate compatible CAD and can only view current design information from other offices (i.e. that which is most well developed and at the highest approval status).

Responses should always be given promptly and revisions minimised. Any problem must be discussed at the first opportunity with all the parties who might be concerned. In practice, all work cannot proceed simultaneously, so submissions should be registered and compared to the design programme to assess their priority and allocate a 'chase by' date.

Liaison with the client would generally be through the lead designer. Liaison with statutory authorities may be through the design discipline most directly associated with the particular area of design work, or through the main or managing contractor.

The comments of the professional design team should be routed back to each specialist contractor via the lead designer and the main or managing contractor so that these have complete knowledge of progress and outstanding requirements for information and co-ordination.

If design work by sub-contractors is to be substantial, it may be worthwhile to agree dates when their information should:

- First be submitted for comment and possible approval.
- Achieve status B.
- Achieve status A.

Where design offices are connected by a computer network or electronic mail, the status of information has to be strictly controlled. In principle, this technology allows information to be processed much more quickly, but the use of information technology introduces a different area of expertise that has to be managed in close liaison with the established tasks of comparing, checking and approving design information for construction. It can give scope for new complications, as well as advantages.

Managers should note that the speed of communication and processing offered by the new information technologies can give rise to a supposition that staff should also work more quickly. If managers encourage staff to believe that speed is all-important, some staff may pay insufficient care and attention to the checking of the inputs to their work, their method, including liaison with others, and to the checking of the outputs. As was true in the past, more haste can still lead to less speed. Information technology can be used to increase the thoroughness and precision of design work and its capability to work faster than pen, paper and calculator should be taken as an opportunity to improve working conditions and to allow staff to be pressured no more than is best for their individual performance.

Notwithstanding the technology, it can be worthwhile allocating staff time specifically for the purpose of checking the co-ordination of designs. On large projects this may be a full-time post over the periods of the most intense design output.

7.21 DESIGN DECISIONS 'JUST IN TIME'

As mentioned previously (for example in section 6.3, page 90), there may be a number of reasons why a design decision is deferred:

- A client may wish to hold some design choices until the latest possible moment to make sure that they have the latest equipment or fashion.
- Tenants or users of buildings may not be identified until a late stage.
- In shopping centres, the division of areas into shop units may depend on a retailers' wish to rent space at quite short notice, or an office building may not be sub-divided until the tenant has signed the leases, considered how best to arrange their work-flow and staff and decided on the locations for partitions.
- The tenants of offices, shops and factory units may alter the layout of toilets and other facilities at the last minute.
- Decisions about finishes such as colours, carpets and fitments may be postponed to the last minute in any building type.
- It may even be that a master craftsman is allowed scope to finalise some details as he is working on site.

The items listed above can often be grouped into a fitting-out contract with a separate time-scale to the main construction contract. Where this is not possible, uncertainties can be managed as outstanding information, with changes rigorously controlled by the design team leader in liaison with the contract managers.

The risks of delay to completion must be carefully considered. Risk analysis is a specialised area and it would be wise to ensure that the managers who are responsible for such decisions have at least a rudimentary training in this subject, for example to appreciate how apparently unlikely impediments can combine with one another to result in a surprisingly high probability that work will be delayed.

For fitting-out and the other sub-contracts, or packages, where decisions may be delayed, it is essential that the design team leader and sub-contractors know:

- Which design decisions can and which cannot be left until the last moment.
- The logistics of the site operations that determine the start and finish dates.
- The design options and the acceptance criteria that may apply.
- The precise information requirements of outstanding decisions such as site dimensions, drawings from sub-contractors working in adjacent areas, samples of work to prove the quality of materials and methods of fixing.
- The resources necessary to complete the design and whether these have been procured and scheduled on the programme.

- The identity of those who must chase, check and authorise outstanding design information, the procedures to be used and how long these will take.
- The periods that apply for ordering, designing, manufacturing and delivering materials and components.
- The risks of delay in the preparation and authorisation of outstanding design information, together with comparable risks of delay while goods and services are on order and during work on site.

Leading designers who work on packages with 'just in time' decisions should be informed about the risk factors, such as those listed above, so that the design team as a whole co-operates in minimising potentially adverse affects.

It is possible that remedial work may be needed on site to accommodate last minute design work, including operations such as cutting holes, or scabbling-back over-sized concrete, and it is easy to overlook factors like this in deciding the sequence of site operations. A margin for error in the construction programme is, therefore, indispensable.

SUMMARY

The construction details for any but the smallest of buildings are produced by a number of designers, working in different offices and for different companies. Each of these offices has its own programme of work and it may not be easy for the managers of any particular project to ensure that all the required design information is exchanged between these designers in accordance with the project programme. Special administrative and leadership skills are necessary to ensure that the 'gears' of design production 'mesh' effectively and consistently.

The planning of design work affords a good opportunity to develop communication and co-operation between the dispersed members of the design team. The production of the work programme may be less important than establishing a common understanding of what has to be done. Each specialism should develop an awareness of the part that the others should play, not only in terms of what each should produce, but also in terms of the information that each specialism will require from the others, when, why and in what form.

Many requirements for the exchange of design information can be identified by systematically searching for dependencies between decisions that are to be made in different areas of the design work. As different design specialists are brought into the team, their information needs should be checked and their part of the design production and manufacturing programme should be discussed and agreed. Sometimes this may lead to adjustments to parts of the programme that have already been agreed between the designers who are already in the team.

Iterative procedures for making design decisions may be planned for, whereby members of the design team release provisional information to other specialists, if this is needed, so that they can proceed with their work. At a later stage, details of the design may have to be agreed and finalised. To avoid confusion, the status of design information has to be made clear, especially where it is communicated between different companies or design specialisms. Design information may be provisional, working (i.e. mostly, but not entirely reliable) or approved for a particular purpose, for example, a town planning submission, tendering purposes, manufacture or construction.

Where a long period is needed to prepare for the manufacture and delivery of certain parts of the building, it may be necessary to freeze the related area of the detail design at a relatively early date. This may also require final decisions to be made in contingent areas of the design.

On fast-track projects, many design freeze dates relate to the release of packages of documents for tendering. It is also necessary, irrespective of the method by which construction work is procured, to freeze areas of the design, at various stages of design development, so that following work can be based on solid decisions and therefore carried out efficiently.

Fast-track programmes may make float more freely available, but they may also generate second and third parallel critical paths. Managers should not only pay attention to the potential to speed up the work-flow in this way, but should also be aware of the pressure that this can put upon the design team. Stress can be relieved if the designers are able to exercise influence over their own programme of work, and to liaise informally with other members of the team.

Time has to be allowed for checking and authorising all the design information, before it is issued. This work may be done through a formal process, such as a series of design reviews, to comply with the project's quality management system.

A good system of change control should be set up and operated, to limit the considerable potential that uncontrolled design changes have to wreak havoc to the design programme. Where design decisions are deliberately postponed to a late date, this area of work has to be carefully considered in advance, so that any changes that might follow from such decisions can be implemented within a planned period and without exceeding the agreed cost plan.

Particular attention should be paid to the procedure that is operated to check design information that is generated while construction is in progress. Full cross-checking, with other elements of the design, can take a long time, if information has to travel between several groups of people. Information technology can be used to make this procedure more efficient, but as with any area of innovation, this requires not only good technical skill, but also experienced and effective management.

NOTES AND REFERENCES

1. See, for example, Chapter 9 of Kast, F.; Rosenzwieg, J. 1981 *Organisation and Management – a Systems Contingency Approach* (International Student Edition). McGraw Hill.
2. In the 1990s, the sections of work are likely to correspond to '*A Common Arrangement of Work Sections for Building Works*', co-ordinating Committee for Project Information, Royal Institution of Chartered Surveyors, published in 1987. The main sections are identified by a letter and the sub-sections by a number. A detailed list can be found in recent editions of the Standard Method of Measurement (SMM 7 onwards).

CHAPTER 8

RESOURCES AND PRODUCTIVITY

8.1 INTRODUCTION

Chapter 2 explained that the quality of design output depends largely on the inputs to the work. The amount of time available is one of the key inputs. The period of calendar time available for each work stage can significantly influence the pressure on the designers to work fast and co-operate effectively. The difficulty of achieving quality and time targets will be affected by the number of designers who are assigned to the work, and the experience and skill that each can bring to bear.

Design companies and departments need to measure the progress of work, not only to assess whether they are keeping up with the output schedule, but also to assess whether production is costing more or less than allowed by their fee bid. Specialist contractors and contractors on design and build contracts may be a little less concerned about profits made directly from design work, because they realise profit from the complete commission, which includes the manufacturing, the installation and the building work. In fact, the latter companies may benefit from allowing some overspend during design development, if this enables them to produce the elements of the building more economically.

Progress in design work can be exceptionally difficult to measure and quantify, largely because the completeness of many design tasks is not evident from the computer files or pieces of paper that represent the design, but depends, to a great extent, on the invisible relationships between this output and the acceptance criteria. Among those criteria is the relative perfection of co-ordination between all the fragments of information that make up the model of the intended building. In principle, the use of computers should make it easier to assess the co-ordination of design work. This aspect of computing, however, is still at a relatively early stage in development and the skills necessary to apply it are often not available in practice.

This chapter considers the allocation of designers to design projects and how the pressure on them may vary according to their productivity. The measurement of productivity depends on the measurement of progress, which is attended by the difficulties mentioned above. The pressure on designers may also depend upon what further work there is 'in the pipeline', as they may be expected to complete their present work by a particular date and move on to another project.

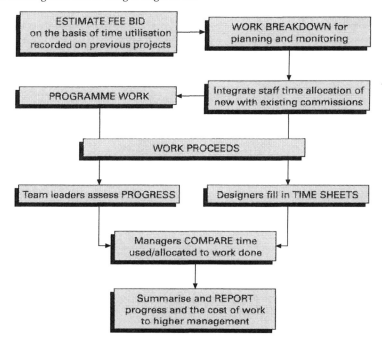

Fig. 8.1   Overview of a resource planning and monitoring system

8.2   OVERVIEW OF PLANNING AND MONITORING

Figure 8.1 indicates the main steps in planning and monitoring work.

The second activity shown, preparing the work breakdown for planning and monitoring, would commence after a design company has been appointed to the commission. The programme of work has to reconcile the requirements of this new commission with the demands of the existing work load on the design staff. Where a new project demands more than the available resources, the programme would have to include recruiting activities. In some cases, staff may need training to meet the needs of the new commission and there may be physical requirements, such as the need to set up a new office and install new computing and communication equipment.

When work is under way, a check is kept on the expenditure of time and money, in comparison with progress in delivering design work. On larger projects, gathering information on progress and designer time use can be a significant task, which itself would figure on the time sheet of the design team leaders. On very large projects, staff may be employed specifically to keep account of progress and prepare reports for consideration by the senior managers of the design practices, the project and the client.

8.3   FEE BIDS AND RESOURCE BUDGETING

Before they are appointed to work on a project, several companies may be
competing to secure each design commission. They judge what fee to bid on the
basis of their experience with previous work. Many companies assume that work on
new projects can be done more efficiently than previous work, because they are
investing in better equipment and their staff will learn how to make more effective
use of their time. By setting challenging but realistic targets in this way, companies
aim to secure more work, in the face of the competition. As each commission
proceeds, their main concern is then to keep ahead of the challenging budgets that
they have set themselves.

Where clients seek competitive fee quotations, design practices may be tempted
to bid on the basis of a minimum estimate of the designer time required to
complete the work. The bid then sets a limit to the designer-hours that can actually
be expended and hence also on the volume and detail of design information that can
be produced.

All competitive tenders are accompanied by the risk that the lowest bidder
will be the company that makes the biggest omission in their estimate! In the
case of design work, any inadequacy may be paid for by the client ten times over,
through difficulties in the construction and use of the building. In the industries
of mass-production, customers can get information about the quality and
effectiveness of a product from previous buyers. The purchasers of building
work can only get information about the inputs to design work, for instance,
by obtaining references from the previous clients of the design companies they
think of employing, but no information exists about the product they will
actually get.

Clients may judge the adequacy of fee bids, to pay for the work that should be
done, by their previous experience, or against guide tables that have been published
by various professional institutions (e.g. RIBA). They may also expect advice from
project managers and client organisations that exist in some specialised sectors,
including public services.

Design fees are often quoted as a percentage of the final construction cost. It has
been suggested that this gives the professional team a vested interest in maximising
the cost of construction, rather than keeping it on target. This tendency is
countered by checking on the past success of the member companies of the design
team at the time when they are selected.

There are two other variables that need to be taken into account. The first is
the type of project, for instance, leisure complexes, office blocks and industrial
buildings would each have different percentage fee. The second is the size of the
project; smaller projects generally require more design input than large ones, in
proportion to the construction cost.

Clients and project managers should ensure that fee bids allow for the full cost
of providing the required quality of service, without omissions. Documentation
that is prepared in order to invite fee bids for design services should call for a

breakdown of the work which will be adequate to demonstrate that sufficient designer time has been allocated. This breakdown would also facilitate the monitoring of progress by senior management, to ensure that the time of the design team is spent where it will be most effective in the client's interests.[1]

8.4   WORK BREAKDOWN STRUCTURE

Part of a work breakdown table setting targets for the use of designers' time is shown in Figure 8.2. and is typical of the type of tables that are needed from the beginning of each stage of design work. This example only expands on RIBA stages E and F and, in practice, it would be further detailed to identify the individual drawings and specifications for each construction work package. In turn, each product deliverable, or information packet, should be described by detailed task lists that outline the steps of production. This can be done at the same time as the work planner is searching for the dependencies of each task on information from the other design work inputs that are required to complete the logic of the design production programme.

Figure 5.5 (page 78) and Figure 6.8 (page 98) include other activities that should be taken into account when allocating the overheads to the design process, in the form of administrative and management staff time.

The lower levels of the work breakdown can be studied informally by the design teams as work progresses. It is better if the principal divisions of the breakdown, as shown in Figure 8.1 (page 128), can be standardised, since this helps with the use of data from past projects in preparing fee bids and organising further work.

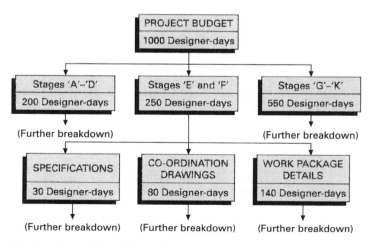

Fig. 8.2   Outline of a work breakdown, focused on the deliverables of one design discipline

8.5 FRAGMENTATION OF RESOURCE MANAGEMENT

Design teams, practices and companies generally consider the management of the time of their particular staff as their own concern and not something that could or should be managed on a cross–company basis. This is largely a matter of practical necessity, but it is also due to the wish of each company, or group within a company, to retain control of their resources to protect their profitability and the survival of their business.

It rarely happens that exactly the same selection of design companies, or teams, is made for successive projects. Since each project organisation is freshly put together, there is no reason to expect that the co-ordination of design work will proceed predictably. It follows that close control has to be maintained over the progress of work, in each team, to ensure that excessive designer-hours are not used.

8.6 RESOURCE ALLOCATION

The estimation of fees responds mainly to those currently charged by competitors, so the utilisation of fees for various products has to be calculated from the bid, rather than the other way round. This is clearly an expert task, that requires detailed knowledge of the designer's craft, the individual designers available and the cash-flow of the particular business.

Figure 8.3 illustrates how target designer-hours, or days or weeks, can be generated for each stage of the design work. The calculation begins from the agreed fee, which is fixed at the outset of the work. This is split into allocated budgets for each stage of the design work and split again into allocations to overheads, target profits and the designers' salaries.

Budget capital cost of construction:	£5 000 000
Proposed total fee for this discipline:	2%
Value of proposed fee:	£100 000
Percentage of fee for detail design:	35%
Value of fee for detail design:	£35 000
Target profit margin:	10%
Therefore:	
Target cost of detail design work:	£31 500
Ratio of overhead to direct designer cost:	1.1/1
Direct designer (target) cost:	£15 000
(i.e. target overhead cost: £16 500)	
Average direct cost per designer hour:	£20
Designer-hours available for detail design:	750
Performance on a previous (comparable) project:	
50 drawings produced in 1250 hours (using CAD)	
i.e. 25 hours/drawing (approved for construction)	
Target number of drawings to be scheduled:	30

Fig. 8.3 Example calculation of designer-hours for detail design work

The direct cost of employing designers; their salaries, insurance contributions, bonuses, etc., is usually outweighed by the indirect costs of overheads and profit. The overheads would include the provision of the design office, complete with the equipment and the materials, plus the administrative staff and the management and the marketing effort of the design practice. The volume of design output which can be produced for the given fee can be calculated from the amount of the designers' time that can be bought and the productivity of the designers, as measured on comparable past projects. This comparison should take into account not only the type of building, but also the way in which procurement was organised and the quality of control.

The bottom line in Figure 8.3 shows a target number of drawings to be produced. In practice, the deliverables would also include preliminary information, such as the brief or feasibility studies, specification documents, schedules, bills of quantity and so forth. So, the allocation of time to required outputs could be considerably more complicated in practice.

The initial judgement about the time requirement can approximate to a statistical view of previous work. The hours needed for an average drawing, specification or other document are derived from the total number of drawings, or documents, produced in the same phase of past projects, divided by the total hours used by the designers for each of these types of output. If the design office fails to keep records of the time spent on producing particular output in the past, current estimates will have to be made on the rather unreliable basis of memory. The estimation and allocation of designer time is likely to be more accurate if the time taken by the range of design tasks is recorded over a period of years and systematically analysed when projects near completion and recorded in the form of a summary.

In reality, there is no such thing as an average drawing. The time required to produce any particular design output (of all types) depends on many factors, including the extent to which the deliverable is dependent on other design work. A good appreciation is needed of the likely need to adjust design outputs, in response to development in other areas of the design. Thus, assessing the number of hours required to produce the drawings, calculations and specification documents that will be needed for a particular project is an expert task. This is likely to be done by a practice partner or associate in consultation with a design team leader, since staff at this level are closely in touch with the problems and possibilities of producing drawings and documents. The equipment the office possesses would be taken into account, and attention paid to the actual staff who would be doing the work.

8.7 TASK LISTS

These are the lowest level of a work breakdown. Their main function is to ensure that essential tasks are identified, prioritised and communicated to the people who should do the work. Task lists are used as a measure of achievement on a week-by-week basis.

Figure 4.9, discussed in Chapter 4 (page 52), is an example task list. Many variations of content and presentation are possible, each of which may help to show:

- The likely rate of progress.
- Times when more staff are needed.
- Times when staff might be under-occupied.

Task lists should be reviewed and updated by the leaders of design groups or individual designers, to decide their priorities and to assign output targets, most likely on a weekly basis. Particular duties may be highlighted in colour on the copies given to the individuals who should carry out the work, or who should supervise it.

Task lists may be long or short, tidy or untidy. They may be particular to a single person or include an entire team. They may be compiled for any job at any level of the organisation. Task lists should be used to generate discussion, gather insight into what has to be done and foster a general alertness to problems.

8.8   MEASURING PRODUCTIVITY

It is important to take regular measurements of the amount of design work that has been done, to ensure that enough time and money remain to complete the work. A measure of productivity is also vital, to show when the design team must increase its efforts to complete work within the cost allowed by the fee.

If a shortfall of productivity is not corrected quickly, it becomes more and more difficult to put it right, as shown in Figure 8.4. The graph lines plot the values from the tables given in the figure. The calculation is quite simple:

- Column 1 shows the total consumption of resources so far, that is the total time used by the designers. In this example, this is assumed to be proportional to the cost of the work, although in practice salary differentials may be significant.
- Column 2 shows the actual progress, where there is the shortfall below the planned productivity of 10 per cent, in the upper table, or 20 per cent, in the lower one.
- Column 3 gives the amount of work that has yet to be done as a percentage of the whole.
- Column 4 computes the required productivity, compared to that originally planned, based on the total required output from column 3, divided by the unused allocation of designer time.
- Column 5 takes into account that productivity so far is only, say, 90 or 80 per cent of that planned. The calculation shows the increase in productivity, above that achieved so far, that would be needed to complete the design project, or stage, within the allocated designer-hours. In this column, 100 per cent represents the actual productivity so far (not the original target).

This calculation serves to show that, if productivity falls, say, 10 per cent behind the planned output, it can be fairly easy to rescue the profitability of a commission,

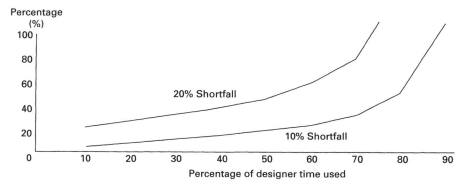

Calculation for the lower curve: 10 per cent shortfall of productivity

Percent designer-days consumed	Percent of work done	Percent of work yet to be done	Required productivity compared to that originally planned	Required increase in productivity above that achieved so far
10	9	91	91/90 = 101.0%	(101.0/.9) − 100 = 12.3%
20	18	82	82/80 = 102.5%	(102.5/.9) − 100 = 13.8%
30	27	73	73/70 = 104.3%	(104.3/.9) − 100 = 15.9%
40	36	64	64/60 = 106.6%	(106.6/.9) − 100 = 18.4%
50	45	55	55/50 = 110.0%	(110.0/.9) − 100 = 22.2%
60	54	46	46/40 = 115.0%	(115.0/.9) − 100 = 27.8%
70	63	37	37/30 = 120.3%	(120.3/.9) − 100 = 33.7%
80	72	28	28/20 = 140.0%	(140.0/.9) − 100 = 55.6%
90	81	19	19/10 = 190.0%	(190.0/.9) − 100 = 111%

Calculation for the upper curve: 20 per cent shortfall of productivity

Percent designer-days consumed	Percent of work done	Percent of work yet to be done	Required productivity compared to that originally planned	Required increase in productivity above that achieved so far
10	8	92	92/90 = 102.2%	(102.2/.8) − 100 = 27.7%
20	16	84	84/80 = 105.0%	(105.0/.8) − 100 = 31.3%
30	24	76	76/70 = 108.6%	(108.6/.8) − 100 = 35.7%
40	32	68	68/60 = 113.3%	(113.3/.8) − 100 = 41.7%
50	40	60	60/50 = 120.0%	(120.0/.8) − 100 = 50.0%
60	48	52	52/40 = 130.0%	(130.0/.8) − 100 = 62.5%
70	56	44	44/30 = 146.7%	(146.7/.8) − 100 = 83.3%
80	64	36	36/20 = 180.0%	(180.0/.8) − 100 = 125%
90	72	28	28/10 = 280.0%	(280.0/.8) − 100 = 250%

Fig. 8.4   Increase in productivity needed to recover from a shortfall in output

provided that action is taken even as late as half way through the work. However, if the shortfall is 20 per cent, or it is not corrected immediately, recovery would be much more difficult.

In practice, some designer time may remain unallocated at the beginning of a project, to provide a cushion against under-productivity. However, this may be a self-delusion, if allocations are made from an inadequate work breakdown and listing of tasks.

8.9   EARNED VALUE

Earned value analysis is an approach to assessing how much payment a consultant or a contractor is entitled to claim as work progresses. During construction, this value is often calculated in terms of the presence of materials and installations at the building site, rather than what the work may actually be worth to the client. If a contractor defaults in some way and another company has to be employed to correct or complete the work, clients would be likely to find that they pay more, in total, than the original contract. This shows that the value to the client of work done can be less than the current valuation. At certain stages of the work, this difference can be quite considerable, without even considering the potential losses that a client could face from late hand-over of a completed building or facility.

The value of design work can be quite difficult to gauge while this work is still in progress. For instance, the detailed designs are only of value to the client if a builder can use this information for construction purposes; if work is cut short before it is complete and co-ordinated, another design practice might charge a heavy fee to finish it. This is because long hours might be needed to fully understand the partly complete work and tease out the inconsistencies before moving on to complete the design deliverables.

In traditional procurement, the value of a design is realised in three stages:

- When planning approval is gained, which often increases the value of a site.
- When tenders for construction are received, showing that the designs can be built within the client's budget.
- When the client realises a profit on the project, through the use or sale of the building.

Where buildings are procured through design and build, or when design information is produced in fast-track packages, the realisation of value may not be achieved at these specific milestones. In any case, the full value of a design may not be appreciated until a building has been in use for many years. The implications of these differences, between earned value and actual value, can be difficult to quantify. Nevertheless, the concept of earned value is useful, even if it is applied somewhat simplistically, to help the managers of design work to keep in mind the implications of any shortfall in productivity.

Conventionally, earned value measures progress in comparison with the work breakdown. If the output at any given point in a programme of design work has taken exactly the allocated designer hours, then the earned value of work is assumed equal to the cost of those hours, plus proportionate allowances for overheads and profit, in accordance with the fee bid.

If costs ever exceed the earned value, profits are threatened. If, at the end of a job, a shortfall in productivity must be met from the profit margin, this may compromise a company's expansion or investment in new equipment. If there is a large shortfall, this would have to be met from the company's reserves, loans or the sale of assets.

The following example illustrates how the implications of a shortfall in productivity can be worked out, at a particular point in a programme of design work. Such bad news would call for urgent remedial action from the design managers.

Example 8.1   Implications of a shortfall in productivity

If the fee for a design phase is £75 000,
and the intended profit is £5 000,
then the planned cost of work is £70 000.

If the actual cost of the work done at a particular stage is £42 000,
and the planned cost of this work (the earned value) was only £35 000,
then the cost exceeds the earned value by the ratio of 1.2:1 (£42 000/£35 000).

Unless the poor performance is improved, the predicted cost of the work in this phase
would be 1.2 times more than planned, i.e. £84 000, and the intended profit will be
wiped out, resulting in an overall loss of £75 000 – £84 000 = –£9 000.

8.10   PESSIMISM AND OPTIMISM IN MEASURING PROGRESS

Progress with any particular task is often estimated as a percentage complete,
but care should be taken to allow for all further work that could be required.
If a designer thinks a sketch, drawing, schedule or specification is 90 per cent
finished, discussion with his team leader may bring out points for correction or
improvement that reduce this estimate to 75 per cent complete. When the work is
presented to others, such as the client, other design disciplines, the local authority
and specialist suppliers, some of it may be immediately accepted, but some may
remain under discussion, requiring additional unscheduled design tasks to be done
along the way.

Measurements of progress therefore need to take into account:

- The completeness of task lists. Many tasks may never be mentioned on the work
  breakdown, so, in assessing progress, or earned value, a judgement must be made
  about the percentage of the total work load that is scheduled.
- The person assessing the progress, who may be a pessimist, realist or an optimist.
- The quality of the work. Progress should only be claimed for work in the
  proportion of how well the designs conform to the requirements as represented
  by the brief and the performance specifications that expand on it.
- The co-ordination. If some designs do not dove-tail well with those developed so
  far by other designers, any claim of progress made for them must be challenged,
  even if the designer hopes that the other designs will be adjusted to suit.

8.11   SETTING-OUT PROGRESS CALCULATIONS

A method of estimating the overall completeness of a design stage or phase is
shown in Figure 8.5. The figures given for the days needed and the progress are
arbitrary and are only intended to illustrate the method.

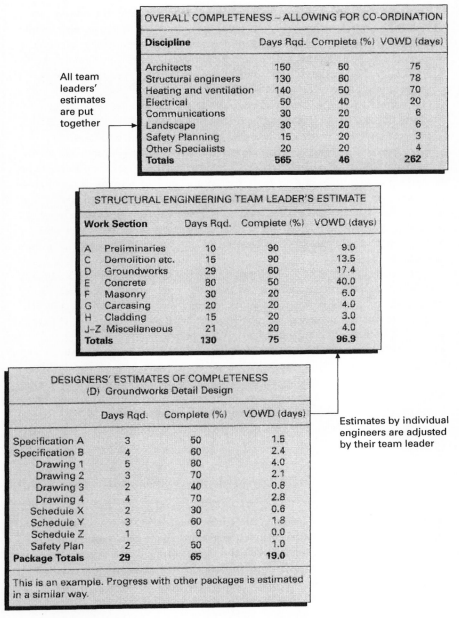

All team leaders' estimates are put together

OVERALL COMPLETENESS – ALLOWING FOR CO-ORDINATION

Discipline	Days Rqd.	Complete (%)	VOWD (days)
Architects	150	50	75
Structural engineers	130	60	78
Heating and ventilation	140	50	70
Electrical	50	40	20
Communications	30	20	6
Landscape	30	20	6
Safety Planning	15	20	3
Other Specialists	20	20	4
Totals	**565**	**46**	**262**

STRUCTURAL ENGINEERING TEAM LEADER'S ESTIMATE

Work Section		Days Rqd.	Complete (%)	VOWD (days)
A	Preliminaries	10	90	9.0
C	Demolition etc.	15	90	13.5
D	Groundworks	29	60	17.4
E	Concrete	80	50	40.0
F	Masonry	30	20	6.0
G	Carcasing	20	20	4.0
H	Cladding	15	20	3.0
J–Z	Miscellaneous	21	20	4.0
Totals		**130**	**75**	**96.9**

DESIGNERS' ESTIMATES OF COMPLETENESS
(D) Groundworks Detail Design

	Days Rqd.	Complete (%)	VOWD (days)
Specification A	3	50	1.5
Specification B	4	60	2.4
Drawing 1	5	80	4.0
Drawing 2	3	70	2.1
Drawing 3	2	40	0.8
Drawing 4	4	70	2.8
Schedule X	2	30	0.6
Schedule Y	3	60	1.8
Schedule Z	1	0	0.0
Safety Plan	2	50	1.0
Package Totals	**29**	**65**	**19.0**

Estimates by individual engineers are adjusted by their team leader

This is an example. Progress with other packages is estimated in a similar way.

Fig. 8.5 Example 'bottom–up' calculation of progress in design work

Estimates of progress are generally made and reported 'from the bottom, up', as shown in Figure 8.5, which takes one discipline, structural design, as an example. The engineers would review their list of deliverables for each package and one set of estimates is shown in the bottom box, for the detail design of groundworks. VOWD is an abbreviation for the value of work done and represents the sum of the progress on each item of design information.

In the middle box, the structural engineering team leader has revised the estimates of his staff. In this example, the groundworks package is shown as 60 per cent complete, 5 per cent less than the estimate of the designers.

The top box shows that a similar downward adjustment has been made by the design team leader, or project manager, when the degree of co-ordination between the outputs of the different disciplines has been taken into account. It is quite important that senior managers should discuss such adjustments with each discipline, so that they are not overoptimistic about their progress. Otherwise, levels of productivity could fall below what is needed to complete design work on time, which could lead to the allocations of staff, budgeted by each discipline, having to be exceeded.

Appropriate adjustments for each project, discipline and stage of work can only be learned from experience and it is not possible to give specific guidance. To illustrate the principle, however, it can be assumed that, at the design team level, one per cent may be deducted for each 10 per cent of outstanding work. Thus, an estimate of 35 per cent complete might be interpreted as no more than $35 - (65/10)$ per cent, which equals 28.5 per cent complete. A greater adjustment may be needed if a substantial proportion of the work has not been clearly defined on schedules, for instance, if only deliverables are shown, a lot of preliminary work may not be sufficiently represented. A similar 'rule of thumb' may be applied a second time, by the design team leader, or project manager, as explained above. It may happen in practice that the designer's view of his progress is more than 100 per cent over optimistic!

8.12 TIME SHEETS

To assess progress and productivity, designers must record their use of time against activities that have been planned in advance. To ensure that the designers are sufficiently aware of the definition of planned tasks, team leaders and managers should discuss the work breakdown with them in some detail. This does not preclude the revision of task lists as work progresses, which commonly occurs.

Most time sheets have columns or codes that differentiate the project stages and the type of work on which time is spent. The specific drawing or document reference may also be recorded. This helps when totalling the figures at the end of each work stage and project. Figures are collated in each company to assist in identifying unproductive time and to improve estimates for future fee bids.

8.13 CHECKING AND DESIGN REVIEW

These are both needed, not only to ensure that design information meets requirements, but also to validate measurements of progress. The principal distinction between them is that checking should be continuous, whereas design reviews occur as an event.

Checking involves an ongoing examination of the outputs of both the completed areas of the design and those that are incomplete, to ensure that they are consistent and conform to the statutory regulations and codes of practice in addition to the client's requirements. It may be useful if an experienced person from outside the design team is able to assist with the checking process and the design reviews, to contribute an objective view, although this is not common in practice.

Some of the computer programs used in design work are very sophisticated. These generally require skill and experience in use and they do not replace the need for the human eye and imagination. The checking of the preliminary and the final design output requires the comparison of sets of information which may be in different forms, such as drawings, calculations, specifications and acceptance criteria. Also, as mentioned previously, interactions between the parts of the building include aspects which may normally be appreciated only subjectively (see section 3.5, page 27), or as systems (see section 5.16, page 77), which are beyond present day computing capability. Conscientious checks are likely to reveal inconsistencies, omissions and faults and, in so doing, they identify areas that need further work. This must be understood when measuring progress, especially in the early stages of work, when task lists, that detail the work breakdown, may have to be extended to include unforeseen requirements for design studies and the communication of design information (between designers and to building contractors).

8.14 POOR PRODUCTIVITY – CONSEQUENCES AND REMEDIES

Insufficient productivity by one design team is likely to place an unfair burden of design work on others. If a shortfall in productivity is not addressed early enough, it may be necessary to 'guillotine' (cut short) the production of design information, so that contracts can be let in accordance with the construction timetable. This can have the effect of passing on unresolved problems to the contractors. This may increase the construction cost, or, if the contractors failed to recognise the inadequacies in the design work, their work might become unprofitable.

Similarly, while design work is in progress, the best design team leaders can only abide closely to their time targets if the designers in the other disciplines also stick to theirs. Design team leaders and project managers should be especially alert to recognise if any member of the team is holding up the others.

The options for correcting poor productivity include:

1. Review the list of required outputs, with a view to removing any deliverables that are not strictly necessary. Such adjustments should be agreed with other design disciplines in case their work could be made more difficult by such omissions.
2. Adjust the design team, for example, by bringing highly productive designers onto the project to set a fast pace for the others.
3. Ask designers to work unpaid overtime, if this is understood to be a normal condition of their employment.
4. Review working methods, although this may not bring any quick improvement.
5. The partners and associates of design practices could bail-out a problem project by working additional hours, along with the design team, for no extra payment. This is a variant on the third option above, which may help to generate a good team spirit (depending on the compatibility of personalities in the team).

8.15 MULTI-PROJECT WORK ENVIRONMENT

It is in the nature of their work that designers will move from one aspect of a design to another in order to create integrated solutions. If they are also required to swap between different projects, the detailed requirements, constraints and partly complete solutions that they carry in their heads, for at least one of these designs, may be disturbed. This loss of both the broad view and the current task focus can be costly to overcome. It is, therefore, important to ensure that, not only are the work schedules carefully thought out and observed, but also that the schedules for overlapping projects are carefully integrated.

If one design team is delayed by another, or if the work is rescheduled for any reason, the status of the incomplete design work should be carefully noted and communicated to other disciplines. A failure to do this could lead to misleading information being left on file or even passed to other designers. Furthermore, if a designer has to set an unfinished task aside, time may also be lost when he resumes it, as he will need to regather relevant information, and refresh the logic of his design strategy. On returning to an interrupted task, designers should be try to be alert to errors and omissions, which may otherwise be easily overlooked and lead to extensive confusion if they are not remedied without delay.

8.16 CORE AND FLOATING PERSONNEL

Since jobs cannot be picked up and put down easily, staff should be allocated to a project for an unbroken period. Ideally, all the required disciplines should begin work on a new project more or less at the same time. This enables their briefing, about the design parameters, output requirements, the quality plan and so forth, to be co-ordinated. Depending on the size of a project, a few leading designers in the team might be assigned to the work earlier, so that they gain a good understanding of the commission and can help the main body of members of the team, who join it later, to become familiar with the work quickly and productively.

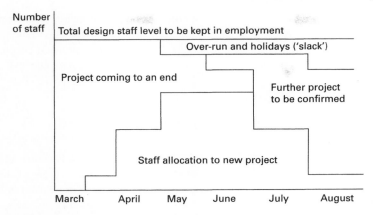

Fig. 8.6 Allocating design staff to successive projects

This leads to the pattern of engagement shown in Figure 8.6 as an interlocking histogram. Members of the design staff are transferred progressively from a project that is finishing to a new one. The picture may be complicated by the need to recruit staff, if there is an increasing company work load. Some of these new staff may need particular training to work alongside established members of staff, so their lead-in to a project might need to be extended.

As the work on any particular design project nears completion, and the size of the team is reduced, it is common for one or two designers to remain with the project longer than the rest, to co-ordinate the design input of the sub-contractors, resolve problems that arise during construction and to put together a file of as-built information that will assist in the management and maintenance of the completed building.

No two projects are identical in their demand for designers, so the staff allocation profiles of successive projects rarely mesh so neatly as Figure 8.6 would suggest. Design practices as a whole will be faced with upward or downward fluctuations in their work load, which have to be met by dynamic staff planning, recruitment and redundancy. It is common for numbers to be topped up with temporary staff from agencies and for design practices to operate as a network, passing overloads between co-operating companies.

Continuity can be created within such a complex work allocation pattern by identifying a few core team members to stay with each project from its start to its completion. The more completely these members can be dedicated to the particular design project, the better the chance that the work will be well co-ordinated (depending on the capabilities and interest of these designers, of course).

Flexibility can also be achieved by assigning some design staff to floating duties, where they stop gaps in projects that may be slipping behind their schedule, or to take over where staff must be transferred to new projects. Since these floating designers are required to adapt quickly to different projects, they should be of above average capability. Sometimes, senior staff are happy to work in this way, but there is no reason why these designers should not be staff drawn from a known agency or from other firms in a co-operating network. They may be thought of as a task force, ready to help to resolve production problems. Floating and core-member roles can be rotated between permanent staff to offer varied experience and interesting work.

8.17   ANALYSING PROGRESS – FOR REPORTS TO MANAGEMENT

Extrapolations of productivity to date and the estimation of trends enable valid comparisons to be made between the relative progress and productivity of different design teams. When this information is presented in reports it can form the basis for decisions, including where to assign resources so that deadlines for each deliverable will be met.

8.17.1   Extrapolations and trends

The difference between these two terms is not widely understood. An extrapolation extends a current direction to some future point. For example, if resources are being used up at the rate of 100 designer-hours per week and 500 hours remain unspent from the budget, then the resources will run out in five weeks.

The concept of trend is more subtle. The rate of change is examined, to calculate what the effect would be if this rate of change were to persist. Considering Example 8.1 in section 8.9, page 136, productivity might be improving, possibly because a good leading designer had been brought into the team. This would make the extrapolation, which suggested a £9000 loss, look rather pessimistic, compared to the trend. However, if in this example, the design team was getting increasingly confused and depressed about the project, the trend might be towards even lower productivity, suggesting the prospect of an even bigger loss.

Neither extrapolations nor trends are established facts, but they can be strong indicators of where action may be needed, to ensure that the business aims are achieved, or, in some cases, to minimise damage.

8.17.2   Analysing progress against the calendar

Progress data should be summarised and analysed, so that managers can assess the need for action or intervention in the organisation of work. Figure 8.7 compares progress data from four projects, referred to as I, II, III and IV. The letters (a) to (p), at the beginning of each line, have nothing to do with work stages, but are labels that are used to show how the numbers are calculated. The approach shown here can be used equally well to report the progress of sub-projects, or discrete commissions.

Figure 8.7 is divided between the point (f) and the point (g). Information above the line refers to progress against the calendar. This may be very different from progress in relation to the expenditure of resources, which is shown below the line.

STEP IN CALCULATION	Projects			
	I	II	III	IV
CALENDAR PROGRESS				
(units = calendar weeks)				
a) Planned project duration	26	20	36	13
b) Present week number	8	16	20	7
c) Previously ahead/behind	1	1	3	1
d) Weeks late (+) or early (–) now	2	–1	3	3
e) Rough extrapolation to finish	6	–1	5	6
(using formula (a) * (d)/(b), so negative is early)				
f) Rough indication of time trend	1	–2	0	n/a
(gauged by comparing progress at several past dates)				
PRODUCTIVITY				
(units = designer-days)				
g) Total budget for each project	300	400	500	100
h) Days used so far	50	370	240	70
i) Value of work done (VOWD)	40	330	210	80
(Base Plan budget)	13%	83%	42%	80%
j) Extrapolated total hours	375	448	571	87
(formula: (g) * (h)/(i))				
k) Projected over/under-spend	75	48	71	–13
(formula (i) – (g), making over-spend positive, under-spend negative)				
l) Previous cumulative productivity	85%	80%	90%	100%
(calculated for the previous report)				
m) Present cumulative productivity	80%	89%	88%	114%
(value of work done (i)/(h) days used)				
n) Productivity trend	–5%	+9%	–2%	+14%
(present productivity (m) – (l) previous productivity)				
o) Required productivity	104%	233%	112%	67%
(budgeted work to do (g) – (j)/(g) – (i) budgeted days remaining)				
p) Required change in productivity	30%	160%	27%	n/a
(required productivity (o)/present productivity (m), less 100%)				

Fig. 8.7 Progress report on a portfolio of projects (equally applicable to sub–projects)

The numbers are derived as follows:

(a)/(b) The planned duration (a) and present week number (b) are usually read from the base plan agreed for the work. An updated version would only be used if changes are so great that comparison against the original plan does not result in meaningful reports.

(c) This is the amount of time that work was ahead or behind, measured previously. This reiterates the report of progress made in the previous regular review and would be calculated as described in sections 4.15 and 4.16 (page 51, page 54). Note that this figure relates to the planned intensity of working. At the beginning of a project, only one or two people might be busy and a two-week slippage is only two designer-weeks. Later on, when five designers are busy, a two-week slippage represents ten designer-weeks and is more serious.

(d)   Weeks late or early, now. This gives the current state of progress, where plus (+) is ahead of schedule and minus (-) means that the work is behind.

(e)   Rough extrapolation to finish. This predicts how early or late the project, or sub-project, might be completed, if the current rate of progress is maintained. This figure is not an accurate prediction, since the planned rate of progress in design work is rarely constant and actual progress will be changed by managerial intervention. Nevertheless, it reflects how well the particular design team is applying itself to the work and slippage may suggest a need to look at areas such as interpersonal relationships, skills and time sheets to find out where the problem is. Of course, it may be that the work programme is at fault and needs revision. If this has already been revised, a systematic difference between the base plan and progress measurements could be expected.

(f)   Rough indication of the time trend. This has limited predictive value, but it should suggest whether progress is slowing down or speeding up and hint at the success or failure of management action so far, in holding production to its programme.

8.17.3   Analysing resource expenditure

In Figure 8.7, the analysis below the line is based on designer-days, or designer-weeks, because the design team leader and his managers are likely to think in these units.

The headings cover:

(g)   Planned budget cost, expressed in the allocation of designer time. This would be derived from the fee bid, after the deduction of direct and indirect overheads and the profit margin.

(h)   Designer-days actually used so far. This derives from time sheet data, which should be as recent as possible, or corrective action may be taken too late.

(i)   Value of work done. This gives the sum of the percentage of progress in all the tasks, multiplied by the designer-days allocated to each, as described in Figure 8.5 (page 137), for the particular discipline.

(j)   The total hours are extrapolated, to give an impression of the implications of the (cumulative) productivity so far.

(k)   The projected overspend, or underspend is deduced from the extrapolation (j).

(l)   The previous productivity is measured from the start of the job, or design phase. This reiterates a figure given in a previous report, as calculated as at line (m). This cumulative figure can obscure recent performance, which may be better or worse than that shown.

(m)/(n)   These are reasonably self-explanatory.

(o)   The required productivity and the change sought to productivity, (p), are calculated as shown in Figure 8.4 (page 134).

8.17.4 Interpreting extrapolations

Project by project, the significance of the progress report shown in Figure 8.7 is as follows:

Project I

This project is behind schedule (d), but the time trend is not bad (f). The staff are insufficiently productive (l) and getting worse (m). However, this is quite early in the project and the required change in productivity is not beyond hope (p).

Action: Since the trends (f) and (n) are poor, a change of manager might be considered. However, one possible explanation of the time trend is that the early design work was being done very carefully. Productivity could now be poor because new members are being brought into the team. If this were the case, then the team leader may be working competently and quite capable of completing the job within both the time and cost targets.

Project II

This project was a little behind schedule (c) but has been pulled around (d) so that it may well finish ahead of schedule (e). However, productivity has been low (l) and although it is improving (m), most of the budget has been used up (h) and the cost will be greater than budgeted (g). Productivity is increasing (n), but the change to productivity needed to protect profits (o) is unrealisable.

Action: The current trends are good, so keep the present manager and accept the marginal overspend. (Perhaps the manager on this job was changed earlier, to pull it around.) Note also that the importance of slippage has to be reckoned in relation to how valued the client may be, relationships with other companies on the design team, the marketing potential of the project etc. This may not show in the numbers that are calculated for the progress report.

Project III

Action to redress slippage (d) has so far only held the previous poor position (c) and late completion is anticipated (e). The projected overspend (k) is substantial. Productivity is inadequate (l) and slipping (m).

Action: Only a small change in productivity is needed to recover the situation, as shown at (p), and this should be achievable if the present team is encouraged and supported by its manager.

Project IV

Such figures are not untypical of an early design phase. Time extrapolations and trends look poor (c, d, e and f), but this might be because work started late or communications with the client were slow. Productivity was good (l) and is getting better (m and n). No figure is targeted for future productivity, at (p), because the designers are not being encouraged to relax! The column records a success for this work group and no action is needed.

SUMMARY

Time is a critical input to design work, which can have direct effects on the quality of work. However, design practices cannot be too generous with the allocation of designers at any stage of the work, because commissions must realise a profit if they are to remain in business. It is therefore imperative to monitor progress, to ensure that sufficient resources remain, at every stage of work, to complete the commission.

A work breakdown not only facilitates this monitoring of progress, but can also assist clients to judge whether the allocation of time, represented in a fee bid, will be adequate to secure the quality of work that is required. However, the value of time is conditioned by the skill and productivity of the designers. In monitoring, practices and project managers must pay close attention to the degree of co-ordination between design information, represented in various forms.

In collating progress reports, it is often necessary to compensate for views of progress which may be much too optimistic at the operational level. The interpretation of such figures is likely to become more accurate and insightful with experience. The effective preparation of design work plans, and interpretation of progress figures, also depends on knowledge of the individual designers and team leaders, including how much care they give to the early stages of design work.

Small shortfalls in productivity can be made good, even quite late in the time-scale of a design project. However, slow output may make it very difficult for other disciplines, at work on the same project, to meet their production targets and to make a profit on their commission. Since estimates of progress are frequently overoptimistic, it is all too easy to overlook shortfalls in productivity that can quickly become impossible to rectify.

The analysis of earned value conventionally focuses on the budgeted cost of work to a contractor or a consultant, rather than the value of incomplete work to the client. This exposes clients to significant financial risks, albeit temporarily. Progress should be verified periodically, for example, through design reviews, attended by the full design team and a representative of the client.

Design managers aim to keep the cost of production within that planned at all times. The cost of production equates to its value as shown on the resource allocation of the base plan. Extrapolations of productivity to date, together with the estimation of trends towards increased or decreased productivity, indicate where management intervention is needed to ensure that the business aims of a commission are realised. Extrapolation and the calculation of trends should follow a defined method, and be presented consistently in reports, so that valid comparisons can be made between the relative progress and productivity of different design teams. This enables senior managers (of the projects and the practices) to form a view of priorities and give leadership, or assign resources, as necessary to achieve their business aims and the deadlines for each deliverable. It is important to appreciate that interpretation of these reports depends very much on knowledge of 'the facts on the ground'. Progress reports should therefore be discussed with design team leaders before action is decided upon.

NOTE

1. One detailed breakdown that may be of particular interest is *Project Team Guidelines – Fee Negotiations and Harmonised Plans of Work*, written for the Association of Consultant Architects and published by The Quadrangle Press in 1988. This was developed from the book by Ray Moxley, entitled *The Architects' Guide to Fee Negotiations*, published by Architecture and Building Practice Guides Ltd., 1984.

CHAPTER 9

IMPLEMENTATION

9.1 INTRODUCTION

This chapter examines the process of introducing or improving a planning and monitoring system for design work in the construction industry. Whether this is for a company or to control design production for a specific project, there are many factors that should be taken into account, if such changes are to be successful. It should be noted, however, that every real situation is different, so the analysis given is essentially theoretical and cannot be exhaustive.

To establish reasonable aims and expectations, the need for change should first be assessed, together with the ways in which the possibilities may be constrained, not least by the cost of making changes. A view must be taken of the appropriate level of involvement of the design and administrative staff in a company, or each section of a project design team. Roles should be defined, so that each person who is involved in setting-up or changing the planning and monitoring system knows what he or she and the other people in the team are expected to do. Appropriate computer tools may need to be selected and skill in their use acquired.

The chapter concludes by listing some critical success factors that may apply to introducing or improving the planning and monitoring of design work.

9.2 THE AIMS

Willingness to spend time and money on developing a more systematic approach to planning and monitoring is likely to be influenced by perception and understanding of the business risks that the system is intended to reduce. Such systems tend to be most favoured by people who are familiar with project management disciplines and are accustomed to analysing work in this kind of way.

There are sound reasons for design practices to develop systematic planning and monitoring of their work, whether or not they are encouraged to do so by project managers and clients. The relevant hazards which planning and monitoring systems address can largely be reduced to three:

- Design cost over-run, which would erode profits.
- Quality shortfall and the liability for design failures.
- Late delivery of design information, for construction and other purposes.

These hazards can put design practices at risk directly, through additional costs, or indirectly, by impairing the marketability of their services or by such factors as the tendency to lose good staff to better organised companies. Poor opinions of a design company may be expressed by the builders who construct their designs, by members of other design companies who work with them, or by clients if they are dissatisfied with the resulting building designs or construction performance against time targets (if late design information is implicated).

9.3   SCOPE FOR ACTION

The scope for action may depend on such factors as:

- The adequacy or inadequacy of the planning and monitoring systems already in use.
- The views of the 'prime mover' for change. This is most likely to be a design practice partner, a project manager, a construction manager, or the lead consultant in a project design team. There are a growing number of specialised quality managers in the industry and people in this role may also take this initiative to improve the control over their design production processes.
- Whether a system is needed immediately, to plan and monitor a particular project, or improvements could be implemented over a longer time-scale.
- The size of the organisation which will use the system, whether it is a small company or department, or a more extended multi-functional organisation.
- The scale and complexity of the project, or projects, where the system is to be used, the procurement methods in use and the stage that the projects have reached.
- The willingness of other companies engaged in design work on a project, or projects, to co-operate in the development and use of systematic planning and monitoring.

The scope of intended action should be clearly defined by the prime mover so that those who are to be involved in this are aware that it affects them.

9.4   EXPLORING THE PRIORITIES FOR ACTION

The scope for change will be limited by what is seen, by those involved, to be both possible and worthwhile. Discussions about the need for changes should begin with a review of the existing systems, to establish whether or not these adequately:

- Analyse data from previous projects to improve fee estimates and project planning.

- Accurately estimate fees, apply cost rates and calculate the designer man-hours.
- Provide model plans for each stage of work and procurement system in use.
- Help to define design project activities (including setting-up the organisation, communicating the brief etc.).
- Relate quality control to work planning.
- Support effective liaison between designers, within the company and between companies.
- Measure the progress of the work and assess the value of work done.
- Gather and evaluate data from time sheets.
- Establish a reporting routine and hierarchy of summary reports.
- Support the control of changes to design work as it proceeds.
- Track outstanding information and ensure that it is supplied on time.
- Meet professional obligations and the terms of commissions.

This list should be extended, to include the criteria which potential users of the planning and monitoring system deem to be significant. Since the purpose is to minimise risks, a broader appreciation of risk analysis could make this more effective, not least by the application of imaginative thinking.

9.5   PRACTICAL QUESTIONS OF IMPLEMENTATION

Besides clarifying the purposes of introducing or improving the planning and monitoring of design work, the managers who wish to introduce changes may find it appropriate to ask, in the specific context of their company and its operations:

- Who should be involved in the process of change?
- Where should responsibility be vested?
- What are the steps in setting-up or changing a system?
- When, in a project life cycle, can changes to planning and monitoring be implemented?
- How long should be allowed for change to take place?
- What does it cost to set-up and run a planning and monitoring system?
- Can benefits be quantified, targeted and reviewed?

Managers may be reluctant to risk losing time, money or credibility in attempting to improve administrative systems. If the work plan for a design project had to be altered or abandoned, staff and managers alike may lose confidence in the planning and monitoring methodology. If this were to happen too often, the credibility of the manager responsible for introducing or developing the system may be damaged.

9.5.1   The people involved

Efforts to improve planning and monitoring practices must be supported by all the senior managers in the project or practice, since they will govern the amount of time that will be forthcoming and whether success will be recognised and appreciated. They (or their board of directors) may initiate action by circulating a bold statement of policy, to outline the need for changes, their aims and how this will affect staff, partner companies and clients.

At the strategic level, it is recommended that particular design teams or projects are targeted and pilot schemes set-up as testing grounds for improved systems. The tactics of the particular programme of improvement have then to be worked out, with specific goals and named individuals taking operational responsibilities. The outcome of the pilot planning and monitoring system may then be evaluated and discussed with other staff, before the system is to be implemented across the company.

As discussed in Chapter 5, there are different levels of planning:

- *Strategic plans* address a project as a whole, or the mid-term future of a business, whether it be the client or the design practice, to target markets, to diversify the products and services, to modify the organisation and to obtain financial support.
- *Tactical plans* explore the more detailed objectives and options that will provide the means of attaining the strategic aims.
- *Operational plans and day-to-day planning* analyse the work breakdown and list the tasks needed to set performance targets.

These levels represent different viewpoints. The attitudes, priorities, modes of thinking and techniques that go with them are not the same. For example, senior managers of design practices and clients may focus on contracts and plan and monitor projects with simple bar charts. Tacticians on the other hand, who may include contract administrators and project managers, set-up the management systems needed for design and procurement, including quality programmes and planning and monitoring systems. To achieve work of quality, with limited resources, operational managers may focus on the goods and services interfaces between companies and work directly with design staff.

The primary purposes of decisions made at strategic level must be communicated right through to operational levels. Tacticians need to pass information up to the strategic planners about the real operational restraints of time and resources. The need for communication and co-operation between the three levels means that, in general, the tacticians are best placed to set-up and run planning systems, that is those who are focused by recent experience of running jobs.

9.5.2   Levels of involvement

A company, or department, may plan and monitor its own work, or it may do this for several companies, or departments. The former is an internal management function and the latter is a project management service most usually undertaken for a fee.

Depending on the size of a design company, the responsibility for planning work and monitoring progress is likely to be shared between the partners, or senior managers, an office manager and the design team leaders. In-house planning and monitoring is generally done through a practice administration system.

Where planning and monitoring is undertaken as a distinct project management function, it is largely of a co-ordinating nature, which may not reduce the work of in-house planning and monitoring. A project manager may have to limit his involvement with small projects, in order to keep the overhead cost down. In such small projects, he or she may act primarily as a facilitator for the planning functions and assist with analysing a limited amount of progress information for routine discussion with the design team leaders and the client.

Large projects may be serviced by a project support office, or PSO, which employs specialists in various areas of management, such as risk analysis, contract administration and quality management, in addition to work planning and monitoring. Care should be taken to ensure that the duties of the PSO do not abrogate those of other companies that are contracted to work on the project. For example, a PSO would not normally measure progress because this can only be done reliably by people who appreciate the degree of co-ordination that exists between different elements of the developing design and this awareness belongs to the design experts.

It is not unusual for design companies to have their own project support offices. The role of such units may depend on whether or not the company is appointed as the project manager for particular projects. If a company has no contract to manage a project, it may still use the in-house project support office to plan and monitor its own work. In either case, this specialised function should not be allowed to take control away from either the design team leaders or the individual designers. The latter understand the requirements of their work in depth, and should retain responsibility for co-ordinating this activity with other design specialisms and delivering the completed work in accordance with a schedule to which they have agreed.

9.5.3   The time-scale of change

This will depend primarily upon whether the planning and monitoring systems are to be introduced for a particular project, developed for a practice, or set-up as a self-standing project support office which can be marketed as an independent service to other companies. It may also depend on the particular pressures of the business market in which the design practice or project management consultant is working.

An evolutionary approach allows time for changes in priorities and procedures to become internalised by staff and hence applied more reliably. This might be appropriate if the company is developing systems for the practice and looks for results over years, rather than weeks. However, if the changes take too long, patterns of behaviour may revert to the earlier practice.

In many cases, it may be better to concentrate the development of a planning and monitoring system on specific projects, where ideas can be implemented by a small group of enthusiastic individuals, rather than expect a large number of staff to work within systems that they do not fully appreciate or necessarily agree with. This approach may also avoid the decline in interest that could occur if the development and implementation of the system takes too long.

Where a planning and monitoring system is needed for a project, it should be set-up as early as possible, when other roles, responsibilities and management systems are being established. Simple systems can be discussed and agreed at a single meeting and implemented within a week, but systems for large or organisationally complex projects may need to be designed and tested over a period of months, while staff are sought and training given.

Every substantial project comprises interlinked sub-projects that have particular aims, for example, those that are specific to each design discipline at a particular stage of work. In the case of construction projects, feasibility studies may be considered as a sub-project, outline design development may be regarded as another and detail design as a third. The more complex the work and the organisation, the more sub-divisions may be needed. For each sub-project, the systems and roles that are required to control the work should be established at the start. As far as possible, each design discipline should have an opportunity to comment on the managerial control mechanisms, so that these can be adjusted, if necessary, to address their concerns.

It follows that the implementation of a planning and monitoring system, on the basis of a project, tends to become stretched out in time, as connections are made with the various companies that participate in design development. The principles of change control apply; the system which is set-up at the outset should only be changed if there is a prospect of some real benefit following from the change. In practice, this is a learning process. When experience has been gained and the system is delivering useful information in an efficient way, good reasons for change are less likely to arise.

The design input of specialist contractors may be especially difficult to plan in an integrated way, because these companies tend to be appointed late in the design process. The systems for quality control must be effective at this stage, since independent working tends to be the norm and many of the designers work under pressure to meet construction deadlines. The initiative for integrating their programme of work tends to fall to the construction management team, or the project management team where this has been defined. This team should work with the lead designers to ensure that design programmes of the specialist sub-contractors will produce information that is properly co-ordinated with other designs and the timing of operations on site. The requisite communication schedules should be worked out in advance. This effectively means that the design planning and monitoring system should be complete before any specialist sub-contract that includes design input is placed. However, details of the work plan may be modified when these contractors are taken on board.

Just as risk analysis is relevant to design work, it can also be applied to the implementation and development of a planning and monitoring system. Similar procedures can be followed, using the work breakdown and task network to search for factors that could disrupt the smooth progress of the intended changes. The search for hazards can be carried out as a process, rather than an event, and raised as an agenda item at planning and progress meetings. Contingency plans should be worked out, to address the most likely or disruptive problems that could arise. In some cases, this may allow for a return to previous practices, while a problem is overcome.

Feedback about the success or inadequacies of the system should be gathered and used to ensure that the planning and monitoring of the design work for successive projects is carried out ever more proficiently. The speed of 'cross-fertilisation' of these control methods will depend largely on the availability of the personnel who have gained experience with them and the enthusiasm of the senior managers to implement and develop the systems. It may be appropriate to run feedback forums to ensure that good and bad experiences are shared.

Periodically, the planning and monitoring system must be checked, to see that it is giving accurate information, and the effective functioning of the system ought to be briefly reviewed towards the end of each phase of a project, whilst memories of what has happened are still fresh.

9.5.4 The cost of planning and monitoring

The setting-up costs for planning and monitoring are largely those of staff time. These will vary according to the scale of operation. The example calculation below illustrates how to account for them. The numbers are arbitrary and actual figures could be very different.

Example 9.1 Setting-up costs for planning and monitoring

Recruit a planner	£5 000
Research computing options	£1 000
Purchase computer hardware	£3 000
(including peripherals and back-up facilities)	
Purchase software (network, database etc.)	£2 000
Train key staff (e.g. planner, facilitator)	£2 000
Staff time spent in training and discussion	£3 000
	£16 000

It should be noted that some costs may be paid over a period of time, for instance, computing equipment may be leased. Tax conditions should be considered. Training may be set against budgets for continuing professional development (CPD).

If the setting-up costs are defrayed across construction work worth, say, £5 million, they might represent about 2 per cent of the total multi-disciplinary professional fee, but two-thirds of the potential project management fee!

To follow through with planning and monitoring a design project of the same order, the cost of staff time might be added up as follows:

Example 9.2   Cost of staff time to plan and monitor a design project

Search files of similar past commissions for significant tasks and dependencies	£400
Set up resource libraries and calendars	£100
Input 150 activities to a computer	£150
Initial optimisation of the schedule	£150
Discuss and agree schedules for the detail design using, say, five key persons, excluding travel	£500
Time at meetings or workshops to inform other designers, the client, etc.	£1 000
Gather, analyse and report progress, say 30 per cent of the activities on ten occasions	£500
Ten short progress meetings, say, five persons	£2 200
	£5 000

This might represent a doubling of current expenditure by the design team, on planning and monitoring, from about half to one per cent of the design fees. Most of these activities can, however, be classified as project management overheads, which could be charged separately from the design fee.

The additional sum required to plan and monitor the project's feasibility and outline design stages would be very small. However, the co-ordination of specialist contractors' design work could easily cost a further £5 000 to administer.

9.5.5   Quantifying benefits

The planning and monitoring of staff utilisation is a normal management function, which is essential to the efficient use of resources in every design company. The resources assigned to the time planning and monitoring of projects, however, may be limited because the benefits arising from this activity are not seen to return directly to the company. The benefits are often shared with the other design companies and construction contractors, through smooth work flow. The client is also likely to benefit from this, for instance, through less claims for extra costs arising in the construction stage.

If good planning and monitoring has led to, say, a £60 000 saving on a £5 million scheme, this benefit might be shared in the ratio of 1:2:2 by the design team, the contractors and the client. One design discipline might therefore benefit only by £3 000, in this example. Compared to the costs of planning and monitoring, outlined in the previous section, this return balances the doubling of effort, without returning additional profit.

The benefits of planning and monitoring can be very difficult to recognise and quantify. It is likely, however, that clients are aware of the comparative performance of different design consultants. Those which deliver results efficiently are more likely to obtain further commissions, in recognition of this successful track record.

What is less obvious is that each design company engaged on a project has to co-operate in systematic planning and monitoring, if out-of-sequence working is to be avoided, with the inefficiencies that follow. If one section of the design team fails to identify its information needs, or deliver information when it is needed by others, this can rapidly undo the efficiencies that would otherwise accrue from systematic planning.

To be worthwhile, the value of forward planning should significantly exceed the cost of the muddle it prevents. To prove this may require some research into the cost of muddle on previous work, including any loss of client goodwill. The costs and potential benefits will vary considerably across a practice, a range of projects, the project phases and each design team.

To measure benefits, productivity figures should be calculated before and after changes are made to the planning and monitoring system. Unfortunately, it can be very difficult to estimate the cost of past inefficiency and assign it correctly to causes. Comparisons may be also difficult because of differences in the nature of design projects. Considerable insight and experience may be needed to interpret correctly the figures that may be gathered.

To make valid comparisons, projects should be classified and compared like-with-like, by considering factors such as:

- The size of projects, usually measured in terms of the construction cost.
- The procurement methods and time-scales.
- The complexity of the project organisation, for example, the number of design specialisms that were involved.
- The knowledgeability and helpfulness of clients.
- The relative skill and experience of designers in each project team.
- The construction technology, whether familiar or innovative.
- How well information technology was used in each instance.
- Whether the project was carried out before or after the introduction of formal quality management and the safety planning required by the Construction (Design and Management) Regulations.
- The effectiveness of change control.

Measures of performance should take other factors into account, such as:

- Did the job go smoothly?
- Was design work done to a high standard?
- Were there less problems to resolve on site than usual?
- Is the client satisfied?

Since the benefits of time planning and monitoring are likely to be distributed across several companies, including the client, the costs should also be spread. The provision of a planner would be less than half the cost of running a planning and monitoring system. To employ someone whose time will be dedicated to nothing but work planning might, therefore, require an annual turnover of some £4 million in professional fees! Most practices are too small to achieve this. This is one reason why construction companies are increasingly active in design management: they have a much more substantial turnover and already employ planners. However, the willingness of increasing numbers of clients to pay a separate project management fee may enable more consultants to retain control of design production management, themselves.

In principle, the existence of a project support office could provide the design companies with a target for litigation if planning and monitoring were ineffective, but, in practice, the success these activities depends very much on co-operation from all concerned and responsibility is necessarily shared.

9.5.6 Keeping things simple

Design management should not be a 'multi-headed monster'! The recent introduction of formal quality management and safety planning, may be seen as more than enough extra bureaucracy for design practices to adopt in one decade. The new role of planning supervisor, together with the increasingly frequent appointment of a project manager, enlarge the project organisations, increase the need for communication, and add complexity to legal liabilities and contracts.

The designing of buildings is a highly professional activity and the designers who have been working for many years in the construction industry know how difficult it can be to co-ordinate the input of different specialists thoroughly and they have experience of resolving these problems. Many faults in design output have nothing to do with inadequate planning and monitoring. Design development may be better served by, say, the development of a quality management system.

Enthusiasm for detailed planning, monitoring of progress and tighter control of work may be lacking if the purposes and methods of working are insufficiently discussed with those involved. Planning and monitoring brings together many differing skills and roles and these should be introduced as simply as possible. At the present time, designers are generally unfamiliar with planning concepts and terminology. If a lengthy learning curve is anticipated, effort should be concentrated on specific projects and small groups of staff, where benefits are most likely to be recognised quickly.

Designers have a fund of knowledge about tasks and precedences that efficient planning must exploit. They may also perceive the level at which planning and monitoring can be effective, and where it might get in the way. An example of planning and monitoring being an impedance would be if the designers worked too rigidly to an agreed programme, rather than seeing it as a way of exploring the problems of a production process that has no single, correct solution.

The difficulty of planning a design project largely corresponds to the number of activities, or events, that should be included in the programme. Figure 9.1 illustrates that large, simple projects might need less than 40 events per £-million construction value, whereas smaller, more complex, projects might need as many as a hundred activities in the programme.

The example programme in the Appendix (page 180) has somewhat less than 140 activities in the design programme, not counting the milestones. In practice, many of these activities would be sub-divided for planning purposes, for instance, every

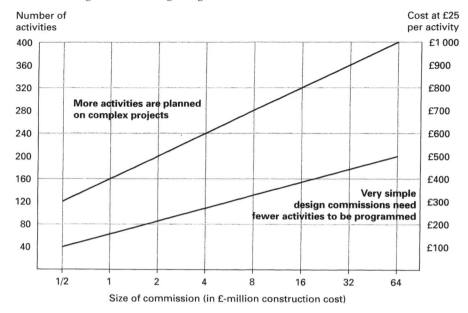

Number of
activities

Cost at £25
per activity

Fig. 9.1   Direct costs of entering and updating design programs on computer

drawing would be scheduled for production at a certain time. This is more a matter of defining the deliverables and how progress will be measured, than modelling the production process by means of the work programme. There would be many transfers of information at the detailed level of design production that should be identified. The more critical ones would be scheduled and the rest managed informally.

Thus, the approach to planning is hierarchical, where the more fixed, formal plan is used as a management tool to target the design outputs and critical information transfers between disciplines, while leaving the detailed organisation of tasks to each company to arrange.

9.6   ORGANISATIONAL FACTORS

Figure 9.2 illustrates, diagrammatically, one way in which the planning and monitoring function can fit in with the established management structure of a design practice. The management functions listed on the left support the work of several teams of designers (represented by the vertical bars). Note that, especially in small companies, several of these functions may be supervised by one manager or practice partner. The design staff report to their team leader, who stands at the head of the bars as indicated by the numbers one to five. The design team leaders are then responsible to the senior partners or managers of the practice and may, themselves, be associates or partners of the practice.

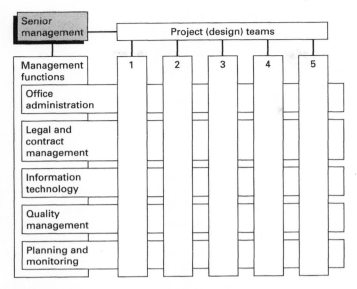

Fig. 9.2   The place of planning and monitoring in a design practice

Planning and monitoring systems belong with quality management as part of the process control function. Both these systems require links to be set up with other companies, or between specialised design departments, to co-ordinate work on particular projects. Changes in management systems may be especially difficult to implement where projects involve many companies, because project managers and design team leaders have little control over how the other companies are organised. The organisation and maintenance of appropriate systems and communications where projects are large or complicated, may therefore be through a project support office as shown in Figure 4.2 (page 39). In this case, planning functions might be supplied by a project management company, under contract directly with the building client.

The involvement of a specialised project support unit should help to ensure that design team leaders and project manager have access to reliable, complete and current information about intended work and actual progress. It may also be of advantage if this view of progress is independent of the team leaders, who may be optimistic or unduly pessimistic.

The objective of a project support office is to inform good management decisions, but it does not take these decisions. It may investigate the work production problem and assist communication by clear presentation of planning proposals, however, plans must be agreed by those who have responsibility for executing the work, because the creation of a planning role does not displace the direct control that design teams have over actual production of designs and design information.

9.7   THE PROJECT OR SUB-PROJECT START-UP

At the beginning of each project or stage, when the teams discuss their programme, each discipline involved should describe the information they require from others and discuss the sequence of work so that a schedule of delivery dates can be agreed. Tasks, such as the production of preliminary drawings and calculations, should be scheduled for use as indicators of productivity at the early stages of design development. Other tasks in the programme may include reaching agreement on the format of drawings, computing protocols, and quality management procedures.

Every designer will appreciate that they need information from the others to get ahead with their own work, but the need to define who needs precisely what and when may have to be stressed. When this information has been represented in planning charts, these must be circulated for study before further discussion and agreement. This approach may be accelerated through workshops, that is, meetings that aim not only to predict and resolve problems, but also to explain the planning and monitoring methods that will be used. Each design specialism should be represented at these workshops, together with the cost consultants and, if possible, the construction companies who will use the design output. Related issues, such as quality management and site safety should be addressed at the same workshops, to ensure that these are fully taken into account in the planning of the design work.

9.8   RESISTANCE TO PLANNING AND MONITORING

The greatest single impediment to implementing effective planning and monitoring may be the common belief that design work cannot be planned in detail. Any lack of enthusiasm for the changes could lead to inadequate work planning, uncontrolled departures from the plan and subsequent disappointment with the results. Designers may resist committing themselves to work sequences because they believe that designs can only develop as a creative response to an internal logic that resides in the design problem itself. Like many strongly held beliefs, this is only partly true. Their help is needed to tease out this logic, so that information transfers between designers can be organised into a logical sequence of key inputs and outputs, which they can understand and appreciate.

It is important to explain and discuss everyone's role fully in the planning and monitoring process, from the humblest worker to the most senior manager. Such explanations and discussions should continue or be repeated until they are understood. They may include references to the importance of company profits as well as just how essential planning is to the achievement of the quality of the work needed to market the company successfully. Anybody who might oppose changes to established planning and monitoring systems should be included in the discussions. This helps to address any reasonable objections they may have and create a climate where ideas are encouraged and the individual feels valued. People support their own ideas more enthusiastically than those imposed by others and broad discussion is generally beneficial in securing active support for change.

Some staff members may be antagonistic to formal planning and monitoring because they fear that it might draw attention to personal shortfalls of performance. Indeed, a planning and monitoring system should draw attention to shortfalls in team productivity, to encourage the instant resolution of problems. People may be more relaxed about this if all the factors that impair productivity are acknowledged, including the inadequacies of the management. If everyone is seeking room for improvement, inadequacies will be seen in a positive light. Openness is to be encouraged, when everyone involved not only learns from their mistakes, but actively supports one another in minimising their consequences.

Senior managers should, therefore, endeavour to:

- Support development of the planning and monitoring system.
- Be ready to explain its mechanisms and benefits.
- Encourage a productive trend at the start of every project.
- Accept that their own shortcomings will be open to scrutiny.
- Be supportive of others, if they have made a misjudgement.
- Set a good example with their own resolve to become better organised.

9.9 A FORUM FOR DISCUSSION

There is a tendency to focus formal meetings and informal conversations on issues of immediate concern, omitting to analyse the full scope of necessary action and develop an efficient method of working. For example, when a client appoints a lead consultant, he would want to delegate control and not become involved in the details of the design production. At the opposite end of the project organisation, designers focus on design problems and do not want to be diverted into detailed discussions about what should be produced two months in the future.

Work planning should, therefore, appear on routine meeting agendas, from the beginning to the end of the project, and at every level in the project organisation. The focus will vary, from strategic concerns about the master programme, to the day-to-day concerns of current task lists. At every level, the facts and purposes of discussion should be thought about in advance and adequate time allocated to them at the meetings.

If topics relating to on-going projects demand all the time that is available at progress meetings, then separate planning meetings may be needed to ensure that sufficient attention is given to this. Other relevant systems may be discussed at planning meetings, such as quality management. While planning and monitoring could be viewed as intrinsic to the quality management system, it cannot be entirely subservient to it, because of the over-riding business consideration that the use of resources has to be controlled in order to protect profits.

9.10   FORMALISING THE PROCESS OF CHANGE

The process of developing a planning and monitoring system should itself be planned, to ensure that aims are precisely targeted and the changes are made within time and cost constraints. It is essential to use staff who have some enthusiasm about management systems, especially at the initial stages.

Agreement should be reached about:

- The primary aims of changes, and what not to change, for instance, the project or practice systems, specific programmes, the methods of allocating work to staff and how progress is measured and reported. The agreed aims should accord with the practice mission statement, if there is one, and with statements of project objectives that may have been prepared by the current clients.
- Specific goals that will contribute to achieving these aims, such as acquiring planning software and learning how to use it, and the value that should be put on each goal.
- The steps required to reach the goals. How the success of each step will be recognised and how each gain will be maintained thereafter.
- A provisional allocation of resources and how their utilisation will be monitored.
- Individual roles in setting-up the system and operating it (see sections 9.11 and 9.12 below).
- The timetable of action and the frequency of reporting on progress.

At any point a lack of knowledge, experience or time may occur. This possibility should be considered from the outset, as part of a risk analysis, and tactics developed to ensure that the development and use of the system will not be hindered. An experienced facilitator may be appointed to assist staff in recognising potential problems and to deal with them.

It is important to analyse not only the established system, but how the proposed system will interact with those of other companies or departments during projects. Therefore, it may be necessary to work with people from these other organisations who have similar job functions and who actually have the authority to adjust those systems.

The progress of changes should be reviewed regularly, not only to improve the approach and to target the next improvement, but also to ensure that people receive credit where it is due. Views of progress should also be sought from the designers and the office administrators, who may be affected by developments in the planning and monitoring system. A list of ideas and suggestions may be set up to record how they think it could be improved.

9.11   THE PLANNER

Unless planning and monitoring is someone's priority, it may not be properly carried out at all times. For this reason, a person has to be nominated as the planner in each company and for each project. Properly considered, this role is really one of being the principal planner, because every person and team in the company or project organisation should be actively planning and monitoring sections of the work.

The nominated planner would specify the appropriate planning methods to be used on a project or phase of work, the kind of information needed for reporting, and the implementation programme. To operate effectively, these people need authority from senior managers to spend time with the design team, or teams, and to have access to relevant data, such as time sheets and progress records. The work programmes should be discussed with the lead designers and passed to the project and practice managers for comment and approval. Project planners should also seek to integrate the work planning of the various design companies and also establish contacts with the construction managers at the earliest possible date.

The planner should liaise with those who will be reporting actual progress, so that their figures relate meaningfully to the plan. Since some individuals will report progress optimistically and others pessimistically, figures may have to be adjusted before they are considered by senior management. The planner should present summary information to management, which the lead designers have helped to interpret. He or she should be ready to reschedule plans for discussion, if work is not being delivered according to the agreed programme.

The planner in one company may have to adjust his or her approach to work, to suit managers and planners in other companies who are working on a particular project. Differences in these approaches may need to be resolved at a higher managerial level.

If an established member of staff is appointed as a part-time planner, in addition to other duties, the time allocated to planning and monitoring should be utilised consistently for this purpose. If this work is not carried out promptly, all the potential benefits of planning and monitoring may be lost.

A design work planner should have a broad understanding of the work done by different design disciplines. He or she should appreciate the way in which design information is produced by sub-contractors as well as the consultants, and understand the use of this information in factories and on the site.

Prime among planning and progress measurement skills are the abilities to:

- Seek information by enquiry and communication.
- Pick out significant factors, by a process of analysis, which includes the application of mathematics and an awareness of management accounting.
- Piece disparate considerations together by a process of synthesis.
- Present questions and conclusions in a way that others can appreciate.

The status accorded to a planner should recognise their need to communicate freely at all levels of the company, when planning and while monitoring the progress of work. This includes contacts as follows:

- With middle managers, who would wish to see efficient operational plans and be well informed to negotiate work targets with their staff.
- With senior managers, so that procurement strategies take into account the work planning implications.
- Client representatives or their appointed project manager, to ensure that design and construction programmes suit the objectives and parameters of the project.

Planners also need to communicate with the people who do the design work and with them:

- Investigate questions about working methods.
- Collate evidence about the dependence between tasks, to identify the critical sequence of operations.
- Calculate the requirements for staffing and the likely time and resources that tasks will consume.
- Work through the implications of progress figures from a variety of optimistic or pessimistic individuals, who may have unequal awareness of the usefulness of the information they provide.

Clear communication requires that the planner should be capable of working within the scope of what others can understand and appreciate, and not attempt to blind people with science or distract them from actually getting work done. Planners need to be aware of developments in information technology and to understand data transfer systems and how CAD and document management systems are used.

Work planners may sometimes be perceived by others as a threat to their control over their own work. This fear can add to the stress of work and some individuals may attempt to return this stress to the planner. Great perseverance may be needed to work consistently under pressure. Resilience is essential, together with solid support from the senior managers. It is important to ensure that the planner commands respect and credibility, so that staff take planning and monitoring seriously and have confidence that benefits will accrue. If a company lacks staff with the necessary skill and experience, a project support office may be brought in, or a planner engaged as an independent consultant.

The designers may be reluctant to cede control to an administrator who favours time and cost as the driving factors in the project or practice. This is more likely to be a problem with architectural work than with engineering, where rational and systematic approaches to design are well established. The planner should, therefore, have sufficient experience of the creative, free-flow of ideas approach to work, used by some designers and a genuine appreciation of its value.

Planners acquire a unique insight into the efficiency and effectiveness of working methods and management styles. Although their position in a company tends to be at one side of the chain of command, experience in this area may be very useful training to move on to a position in senior management.

9.12   OTHER ROLES IN THE WORK PLANNING PROCESS

Other than the principal role of the planner, a further 20 or so roles can be identified in the process of establishing, running and improving a planning and monitoring system. These roles need not each be given to different people. Except for the largest projects or companies, several roles would be vested in one person. On the other hand, where several companies collaborate in design production, the same roles may be given to a number of individuals, working for different companies.

Some of these roles are listed below to stimulate thinking about the many aspects of implementing and running management systems. The system of roles may be quite fluid. Some may fuse where aspects of work diminish in relative significance. Others may split into more specialised functions, for example where several people are working together in a project support office.

9.12.1   Assistance roles

There are a number of people who, by virtue of their particular job within a design team, will have an important role in planning work and monitoring progress.

Design manager/co-ordinator

He or she should be able, not only to direct the design operations, but also to consistently encourage the project team to analyse what they are doing and how they approach their work. The design manager/co-ordinator should assess the actual progress of scheduled work independently of the lead designers in each discipline. Effectively, the role is one of project manager for the design phase.

Lead designers

The lead designer in each discipline must appreciate the purposes and methods of planning and monitoring. They should discuss the task breakdown and programmes of work with other disciplines and their own staff and accept the planned content, form and timing of design output, on their behalf. Subsequently, the lead designer would supply information about the progress of work to his manager and the planner.

Quality manager

This person has a close interest in ensuring that the design process is planned in such a way that the designers will be able to follow the quality control procedures. Relevant input may be obtained from quality managers in each company that will contribute to the design of a building and its construction on site. This group may prove to be key players in the development and use of the planning and monitoring system, in parallel with the generation and application of acceptance criteria for the inputs and outputs of design work.

Practice manager's assistant

This person should regularly distribute and collect time sheets and collate the figures for time expenditure across a practice.

Data analyst and computing technician

Computing knowledge is indispensable to efficient planning and monitoring of all but the simplest project, or the earliest stages of more complex projects. Expert assistance is invaluable in managing data securely and in helping the planner and design managers to use information technology effectively.

Report formatter

Someone who is skilled in presentation is needed to illustrate plans and to clearly demonstrate the significance of the information gleaned from monitoring the progress of the work. This is very valuable in improving communication at meetings and focusing managerial attention. This member of staff should understand what most concerns management, so that they will select the information that will be the most useful to them and present it in an informative way.

9.12.2   Implementation roles

The following roles are relevant when changes are being made to systems of planning and monitoring.

The champion

The authority of a senior manager is needed to insist that the planning and monitoring system, or the enhancement of an existing system, is necessary. This senior manager must be able to allocate staff time and resources, such as office space, computers, software and technical assistance, to the development of the system and will follow through to ensure that they are implemented effectively. He or she should be a capable communicator and sufficiently knowledgeable in this area to explain the concepts of planning and monitoring to the staff, clients and others, together with the related structure of roles and responsibilities.

The champion's deputy

In larger companies, someone may be available to do most of the explaining and pushing on behalf of the champion. He or she should also be a manager who is well versed in the purposes of planning and monitoring and the methods to be employed.

Development engineer or computer systems engineer

The selection of appropriate software, and the hardware on which it will run, is an area of specialised expertise. When a planning system is under development, someone should carefully consider the alternative planning tools and make recommendations as to how the system will work. It may be possible to engage this kind of staff through an IT service company, instead of employing them directly. This latter option has the advantage that back-up personnel may be provided if the development engineer is ill or on holiday.

A data processor

In the programme of starting or upgrading a planning and monitoring system, time must be allowed to generate data for use in testing the computer programs and whether or not the staff can apply the planning and monitoring procedures. This data is most likely to comprise information from real projects that the system should prove capable of handling. The task of generating test data may be straightforward if the records, such as time sheets and drawing registers, have been retained from previous projects. However, effective planning and monitoring should encompass the activities of all the design disciplines. If the requisite historical data is not available from one or more companies, companies on the design team of previous projects may be approached, to borrow test data from their records.

The quality manager

This person (or a quality management group) may help in establishing the acceptance criteria for the planning and monitoring system and using these effectively during its development. They may also give advice in establishing procedures to verify the performance of the system in use and thereby maintain the improvement path of the system.

The facilitator

This person would chair meetings and workshops, both during the process of setting-up a planning and monitoring system and as different groups of the design team or company are introduced to planning and monitoring activities. He or she should be cognizant of the techniques and the types of project where they are to be used. Not least, he or she must also be an excellent communicator and motivator and be able to set others thinking about their options and the potential difficulties.

A 'sounding board'

An intelligent person, who is outside the process of change, should be made available to give an objective view of it. They may comment on presentations about proposed systems and their implementation, if these can be made by the those engaged in this work. He or she may also respond to written proposals. Knowledge of planning and monitoring may not necessarily be important in this particular role, because a person who is not well versed in these areas may more capable of seeing through jargon and challenging other examples of incoherent communication.

A 'fire-fighter'

Someone of authority in a project or practice organisation may be needed to expedite work that has slipped behind schedule or is costing too much, especially if the design manager, or the lead consultant, has become too closely identified with design developments and the consultant team to impose such authority. The champion, mentioned above, may be useful in this capacity, because he or she has a strong interest in the success of the planning methodology and the team responsible for implementing it.

9.12.3   Liaison roles

Designers, their managers and the construction management team have interests in running a work planning and monitoring system. They also have expertise which can assist in setting-up such systems.

The project (or practice) information manager

This person ensures that the status of all information in use is clear and that the designers are working with current information. This minimises the possibility that design might have to be reworked because out-of-date information was used. Note that the scope of this information should be comprehensive, extending to areas such as project handbooks, technical literature, the staff skills inventory, their holiday programme and the content of the planning and monitoring system, as well.

The information technology manager

This role is common in design practices that use CAD. They may inform the work planning staff about the problems and pitfalls of scheduling the use of skilled staff and expensive computing equipment. They may also be familiar with the possibilities of automatic data transfer, in speeding the exchange of design information between companies and thereby speeding the production and co-ordination of designs.

The procurement and construction planner

This person should ensure that lead-times for construction materials or components are fully taken into account, in the detail of design work programmes. He or she might also advise on planning and integrating the design work that is required from specialist contractors.

Contract administrator

People in this role may assist in devising the programme for tendering construction contracts, including the lead-in times required to finalise the content of tender documentation.

Planning supervisor

In monitoring design proposals and drawing attention to their health and safety implications, for the construction processes and use of a building, this person contributes a set of acceptance criteria for the work to be done. He or she should ensure that design work programmes include all the tasks that are necessary to analyse designs from the point of view of health and safety. This would include allowance of time, in the work programme, for contractors to develop and apply their safety plans for manufacturing, delivery and site operations.

The practice manager

Most design practices have managers who prepare the work load projections for the various design teams, to inform project and practice managers about future overload or underload. They might also inform partners about the ongoing profitability of jobs, by comparing information from designers' time sheets with the work planner's assessment of the value of work done. This would focus the partners' attention, in good time, on the jobs that need closer control.

Estimator or scheduler

Fee bidding is an area of special expertise. These estimates set ceilings to the resources that will be available to carry out any commission. This ceiling may make a big difference to whether the work is relatively easy or difficult to do. Companies bidding for design work must also look at the capacity of their office to accept new commissions and this requires close liaison with the management and their work planner.

The marketing manager

Through attention to the marketing potential of particular projects, this individual may influence the interpretation of trends towards profit or loss, in each instance. He or she may also exploit the marketing potential of planning and monitoring, as an aspect of the well-developed management systems of the company or project. This can be very important, where a company is developing their project management capability, as this can secure additional income through separate project management fees.

The training co-ordinator

In each design discipline, someone has to integrate the continuous professional development (CPD) of the staff and balance the skills that are available with those that are needed. The appreciation and application of work planning and monitoring techniques will have a different place in the priorities of each member of staff. The CPD co-ordinator should ensure that they get training in this area which is appropriate to their job and level in the managerial hierarchy.

9.13 ROLE MODELS

Roles are not simply sets of activities for which individuals are made accountable, they are also complicated sets of relationships with other people. Difficulties may be encountered if a new planning and monitoring methodology is introduced without modifying the expectations of all those whose work may be affected by it. Their expectations may be educated, either by employing a professional planner to work with them, or by working with other companies on projects where well-developed systems of work planning and monitoring are already in use.

Much can be learned when a company works in collaboration with others that already have a planning and monitoring system in place. There may, however, be cultural differences in their manager's style of decision making, such as where a consultative approach may not mix well with the more autocratic style of another company. Another cultural matter is the degree to which decisions are considered to be dependent on a search for accurate information, or conversely, reached through judgement which is based on experience and 'gut feeling'. There may also be differences in the attention given to issues such as design quality, production rates, staff morale, profitability, marketing, and innovation, Such differences may be reflected in the work planning and monitoring procedures and how they are applied.

Whether or not experience can be gained through collaboration with companies that already use a planning and monitoring system, design companies may themselves employ an experienced planner to implement a system of their own. This person may serve as an effective role model if their experience is relevant to design work and if they are also capable of communicating to others what is expected from them.

A third way to proceed is by purchasing the experience of planning and monitoring from a specialised company, such as those selling project management services. If skill-transfer is an explicit aim of the commission, this should take account of the extent of the present skills in the design company. Hands–on training should proceed on a live project or series of projects, where the planning consultant would prompt staff to follow appropriate procedures and regularly review and correct the progress made in implementing systems.

9.14 COMPUTING TOOLS

A wide range of software is applicable to planning and monitoring. The selection of which to use, and the hardware on which to run it, depends, to some extent, on the purposes and experience of the user.

Strategic master planning can be carried out with the help of relatively simple tools. The contracts of a project may be programmed simply as a bar chart, because slippage of the major milestones and project phases should be prevented at the tactical level of management. Spreadsheets to calculate the top level breakdown of projected payments may also be relatively uncomplicated.

Operational planning requires the use of a network analysis program.
The temptation to plan in too much detail should be avoided, partly because
fees at the design stage are inadequate to support a detailed analysis. It is
important to recognise just where the sequence of design work is so fluid that the
sequence of work would not benefit from precise preplanning. The key functions
of the planning activity are to provide a sufficient breakdown of work by which
to measure progress and to recognise potential difficulties, such as high-risk nodes,
in advance.

The reporting of progress from the operational level can be assisted by the
integration of spreadsheets into word-processed reports. Some databases provide
powerful reporting facilities, and these may be effective tools for registering and
summarising the progress of predefined design activities. Other network analysis
programs provide integrated functions, to register, analyse and automatically report
progress and the use of staff time.

Most of the network planning systems currently sold are not difficult to learn
and use, although any system may be found to have its particular inadequacies and
faults. Time should be allowed to test the software, moving on to another product
if one is found to be difficult to use or otherwise unsatisfactory.

Data from work programs and progress reports must be backed-up (copied and
the copy stored in a safe place), systematically and at frequent intervals. Recent
back-up copies must be tested regularly to ensure that it is possible to work from
them, if the original data are corrupted. All files should be catalogued systematically
and the unique version of each file that is in current use, referred to as the control
copy, should be clearly labelled as such. Obviously, this label has to be removed or
altered if a control copy is superseded by a later version of the file. This discipline
applies equally to paper filing systems and print-outs should also be very clearly
labelled (often by automatically printing the file name and revision date on every
copy) and indexed. The distribution of information should also be carefully
recorded, so that design managers and the team leaders in each discipline can easily
check that every member of the team is working to the same, current, version of the
work program. (Note that this disciplined approach is also essential in handling the
design information.)

The price of powerful computers has dropped dramatically, but anyone ordering
a particular system should pay due attention to the hardware requirements when
selecting software. Some programs that appear satisfactory in a demonstration, may
perform less well if large amounts of data are fed in. It is important to maintain
sufficient free memory on a back-up computer to take the planning software and
records if the computer that is usually used happens to develop a fault. Back-up
printers should also be available and it is essential that all these machines are
compatible with those normally used for the purpose. Colour monitors and printers
are now standard and available in a wide range of sizes. Printers and plotters must
have an adequate memory capacity, as they can get choked with large amounts of
data, just like a small computer. Monitors should be as large as possible and provide
good image definition, so that the fine print of planning charts and the linking lines
of the network are easily discernible.

Careful consideration should be given to who will use the various pieces of computing equipment, where they will be located and how they and the peripherals will be linked together. All these elements of hardware and any cable connections and switches should also be backed up with readily available substitutes, in case of faults and failures occurring. If money is not available to purchase or lease spare hardware immediate access to replacement machines must be on hand and this should be assured through an arrangement with a dealer, or a service contract.

On large projects, the lead consultant may insist that collaborating companies purchase and become proficient in a particular planning software, but it may be better to use systems already owned and understood by the participating companies, who will include the key contractors and the specialist sub-contractors.

Hardware can be set up and the software installed in a very short time by an expert, whereas this can cause continuing difficulties for a relative novice. Many design practices are still fairly inexperienced in the use of computers and software, so the significance of this point should not be underestimated. Plenty of time should be allowed, not only for staff to learn about the new systems, but also for them to gain experience in applying this knowledge in the office.

Underspending on computer software, equipment, or on expert support, is unwise. Planning and monitoring are at the core of management decision making, and often have to operate quickly. Technical mishaps may be tolerated for minutes, but they can cause serious problems if they last for hours and they should certainly not persist for days. Skimping on advice in this field is a false economy.

9.15   FURTHER NOTES ON NETWORK PLANNING SOFTWARE

There are many competing computer systems on the market and it is difficult to generalise about the key requirements because every planner will develop a slightly different way of working. The considerations given below may be taken into account, however, when software is being evaluated for purchase.

9.15.1   Manuals

Manuals and self-help tutorials for computer programs differ in quality and readability. A good tutorial considerably reduces the time staff are likely to take in learning and applying software. Good help systems (offering advice to the users of software, directly from the computer) are rare, and so a well-written manual may be more useful in answering the questions that inevitably arise when software is put to use.

9.15.2   Chart creation

Planning software varies in the way it enables charts to be created or modified. For example, in setting-up a project plan, activities may be entered into the computer as a list, with precedence also entered as a list. In this case activity codes may have to be considered. Other systems allow the activities to be drawn directly onto a bar chart and show precedence by linking them with lines drawn by using a 'mouse'. Task schedules, set up automatically by the software, can then be viewed on screen or printed.

Interactive systems, that allow the effects of input to be seen immediately, are generally the easiest and quickest to use. In recent years, the cost of powerful computers has dropped so much that the vast amount of calculation needed to run interactive software can be executed with lightning speed on relatively cheap hardware.

Projects, generally, are broken down into numerous activities with the result that planning networks are too big to be viewed in detail on a computer monitor. Some systems can 'zoom' to show the overall pattern on screen, with the loss of some detail. Most systems provide some way to hammock activities, that is, to represent a number of linked activities within one bar of a Gantt chart or node of a network.

Master plans generally comprise simple bar charts. If these are on screen, the detailed activity network for each stage and contract can be brought up by using the hammocking facility. If a contract or stage of work is input first as a bar of fixed length, say, in units of months, and the day to day program of work that fits within this time parameter can be worked out later.

Hammocks can be used to define the activity networks of sub-projects, and it is often convenient to print these on separate sheets of paper as sub-charts. A good system will display on screen exactly what can be printed out on paper. Care must be taken to highlight cross-linkages to activities that are not shown on the particular view or sub-chart, for, otherwise, it can be difficult to trace the logic of a work sequence as it skips from one chart to another.

The work of each design discipline can be shown on a separate sub-chart and the links between these charts represent the key transfers of information between the design disciplines. If it is not possible to show these links clearly, it may be better to include all the disciplines on the same chart. Legibility requires adequate print size, so there is a limit to how much can be shown on one sheet, for example, no more than 30 activities should be displayed on a standard A4 size sheet. Printers that can draw on large sheets may therefore be very useful, where a lot of detail has to be shown on one chart. Small print can sometimes be effectively enlarged by using photocopiers; A4 sheets can be expanded to A3 size in this way.

Large, complex plans can be confusing, whether Gantt or network, and important relationships can be overlooked. To achieve compact charts, some systems enable related activities can be shown on one line. For example, the detail design, measurement, tender and design for the manufacture of one construction package can occupy the same line. It is also helpful if the different work types, or the work of different companies, can be represented by a range of distinctive patterns, rather than colours, unless colour photocopiers are readily available to all disciplines in the design team.

All planning software gives the user alternative views of the work program, as networks or schedules. Many include a database for resource allocation, so that the staff, or teams, are identified along with tasks. This can help managers to recognise where people are over- or under-employed and to modify either the program, the work, or the staff allocation, depending on whether a smooth work flow or a rapid completion is more important at the time.

Many systems permit parts of project plans to be copied or moved, key links to be changed and the views of the network or Gantt chart to be rearranged. Networks used for previous projects can be copied into new networks, in whole or in part. This speeds the setting-up of new ones, but care must be taken to ensure that differences between new and previous projects are recognised and any adjustments made.

9.15.3   Network analysis

The fact that activities can be related by precedence does not mean that they should be managed in this way. Although fast-track construction procurement allows the timing of design and building work to overlap, the master planning chart defines the beginning and ends of key tasks and stages. Network analysis may be indispensable in resolving the detailed logistical problems of meeting each deadline, but this application does not reduce the contractual significance of the simple start and finish dates that may be shown on a bar chart. Networks are invaluable in analysing the necessary, irreducible, duration of critical paths, but the detailed sequence of work may be very flexible, as design decisions and information deliverables are often prepared in the periods designated as float.

Different systems show criticality and float in different ways. This information can make some planning charts look vague or complicated and it is useful to be able to choose whether the display of criticality and float is off or on, and whether or not this information is printed on the paper output.

Colour representation is of undoubted value, where activities are shown that are to be carried out by different design disciplines, or companies. Colour may also be used to highlight the critical path or, in reports, to distinguish the progress to date from unfinished work.

9.15.4   Resource analysis

Resource analysis begins with a resource library. Most systems allow for at least two types of resource, since construction requires labour, plant and materials to be scheduled. In design work, it may be sufficient to schedule the work of people, although it may be wise to include CAD as a separate resource if this presents a bottleneck. Ideally, resources are represented graphically on bar charts and histograms as a pattern or by colour, for example to distinguish the different design disciplines.

Work on tasks may vary in its intensity, through the number of individuals assigned to carry it out. The total designer-days used in a unit period, day or week, can be summed up to show a resource histogram, to indicate whether the plan shows too much or too little work for the available manpower. Ideally, the total of allocated resources is shown interactively on the screen, together with unallocated resources. Some flexibility in the profile of a job should be accepted, as work invariably proceeds at different rates at different times. (The discussion of Figure 4.5, given in section 4.5 page 42, refers to a staccato pattern.)

Resources can be smoothed automatically by some systems. This procedure takes uncritical work from peaks, or periods of overload, and reallocates it to troughs in the programme of work. Since the computer is not able to recognise subtle differences in the priority of activities, output from this facility may not be used without careful scrutiny.

9.15.5   Time and progress

The time units of a project programme deserve attention. Most systems permit charts to be generated showing days, weeks or calendar months. Grids can be selected to divide bar charts into these and other periods such as four week intervals and can be scaled so that they fit the paper when they are printed out. Most systems calculate the duration of work in relation to the calendar, so that weekends and public holidays can be left out of the planning. Several systems allow different resources to have different holidays, which can be useful if the time of individual designers is to be programmed.

Most systems allow current progress to be marked up onto planning charts, usually as percent complete for each activity. Some software can calculate total progress as a sum of the work done on individual tasks, but there is more than one way to do this, so the results need to be considered carefully, at least, until the characteristics of the particular calculation system become familiar. This important information may have to be processed quickly for submission to regular management meetings, so consideration must be given to how quickly and accurately the progress can be recorded and totalled. Clarity of presentation is vital, since important decisions may be based on this information.

Progress may be represented on Gantt charts as an undulating line, to the right of completed tasks, or by shading bars in proportion to the amount of work that has been done. Progress with each task can also be listed on a schedule, with summary totals.

9.15.6   Reporting

There are a number of reporting functions that can be regarded as standard in all network software. These include the automatic production of schedules of activities, which show their start and end dates, float, criticality and committed resources. Some systems also analyse cost and earned value. Alternatively, the reporting may be selective, gathering together the activities of one design discipline, or showing only the activities that are critical or running behind schedule. Planners may choose to transfer such information to a spreadsheet for analysis and presentation.

Mutually dependent design activities may be difficult to show on planning software. These may have to be analysed into activities that are sequentially dependent. The capability of some software to show several activities as bars on one line can therefore be very useful, especially if the linking arrows of Gantt charts remain clear on the output.

To draw attention to high-risk nodes and milestones (which may be one and the same), most planning software provides special predesigned symbols to represent them.

Some planning software provides special reports, or print-outs, on filtered information, showing, for example, only the progress on critical activities, or in one particular discipline, if this has been identified as a distinct resource.

Since the design work on a project is usually shared between different firms and these may use different project management software, planners may often need to integrate the task breakdown and progress information from different systems. Some software allows parts of a work program to be transmitted separately and pasted into the project manager's system, for example, as a sub-chart. Care should be taken, if this is done, to ensure that the links between activities connect up as they should. The planner, or the report formatter, is more likely to review a hard copy of the separate programs and transfer only the key facts onto his or her own master programming chart.

9.15.7   Planning the office work load

Multi-project environments can be accommodated by the more powerful network software. Provided that the system can handle the total number of activities that must be planned and monitored, it is only necessary to set up the work load as if it were one big project. The separate projects of this work load are then set up as sub-projects that share resources, such as the design staff. In this way, the time of every technical or professional member of staff can be planned and monitored on one system. When a project is completed, it can be deleted from the total project and new projects added in as they begin.

The capacity of the planning system should be carefully calculated, to ensure that it can handle a growing volume of data. It is wise to provide much more space than is likely to be needed. The number of levels provided (i.e. hammocks within hammocks) should also be adequate, as the office work load will be divided into projects and may then be sub-divided several times. It is possible that a system, originally intended to handle a small work load, may later be used to integrate the work of several companies on several large projects.

Attention should be paid to how completed projects will be removed from the system. It should be possible to freeze historical data, so that projects can be removed from the system without disturbing, for example, the spare resources viewed in relation to other projects.

9.15.8   Additional information

Invariably, charts are printed with headings identifying the design office and project name, a chart name, date of issue, date of latest revision, standard notes and so forth. Most systems can print additional information on charts, such as the planned start and finish dates of each activity. Some allow notes to be written almost anywhere. Systems that indicate resources by patterns and colours should also print a key to these on the charts to which they apply. This information helps to make the charts easy to identify, use and retrieve when filed.

Some systems allow the description of tasks to be expanded. A notepad may be provided, as a pop-up window on the screen, or activities may be linked to a database where specifications and the like can be developed. Some systems allow notes to be written and word processed directly on the planning chart, and printed together with it.

9.15.9 Saving charts and plans

The development of work plans tends to be a trial and error business. It should be possible to save draft plans and alterations to plans separately from the control charts, to which people are currently working. Base plans should be recorded in write-protected files. It may also be useful to archive charts that plot the progress of the job at various stages as write-protected copies, for future analysis.

SUMMARY

Successful planning and monitoring depend on the skill of managers over a broad spectrum of activities and the flair and technical expertise of designers, who are expected to produce co-ordinated designs within a time-scale. The tools and techniques of planning and monitoring can aid the analysis and discussion of how best to go from 'here' to 'there', but they cannot be a substitute for ingenuity, flexibility and good interpersonal relationships. The latter are essential to create a plan of action, to agree it and to carry it through.

Planning must be done in earnest and, once a programme has been agreed, everyone in the design team should try hard to work to it. Without commitment, the work of different designers can get out of synchronisation very quickly. Anything less than close monitoring of progress can allow the practicality of a plan to degenerate quite rapidly.

Success is reliant on a number of critical factors and, although people will work in a variety of ways, there are some specific points that should always be observed. No doubt, many readers could extend the list given below, from their own experience.

With regard to clients:

- It should be kept in mind that the ultimate purpose of any project is to satisfy the requirements of a client. While service to profitability may be the basic measure of the fitness for purpose of all the management systems, every business depends on client satisfaction, in order to secure further business.
- If a choice has to be made between profitability and client satisfaction, the latter may often emerge as the dominant consideration.

With regard to the design team:

- Staff must be convinced that better planning and monitoring is essential. This requires adequate discussion between all the people involved. The methods and roles must be explained and discussed thoroughly and the team managers must carry others along with their commitment to it.
- The aims and scope of proposed developments to the planning and monitoring system must be made explicit.
- It is essential that anybody who appears to resist changes should be included, in some way, in the discussions about the proposals.
- The advice of staff who devised and operate existing planning and monitoring procedures should not be undervalued, or they may not support the changes.

With regard to the work planners:

- Someone must be responsible for planning and monitoring as their first priority, otherwise it may never receive the immediate and persistent attention it requires.
- The credibility of the planner must be beyond doubt. If the role is not established in the organisation, an experienced planner may be brought in from outside the company.
- The planner should not displace the design team's control over their programme of work.
- The detail of a work plan should only be to a level with which the design team feels comfortable. The plan should be kept simple, while the designer's ability to plan their work in detail should be extended.

With regard to the technology:

- Underspending on computing equipment, software or expert advice can put at risk the success of any planning and monitoring system, to the extent that it relies on this knowledge and technology.

With regard to the organisation:

- Planning and monitoring are part of a system of interdependent management and process control techniques. The development of planning and monitoring methods must be combined with the promotion of a systematic approach to quality management. Attempts to develop one area in isolation from the others may amount to little more than tinkering and the benefits will not measure up to the effort expended.
- Senior staff must demonstrate that they are committed to the work planning culture by trying to work in a well-organised way.
- Planning and monitoring should commence at the very start of a project.
- The project objectives and administrative mechanisms should be explained in a manner that ensures they are understood and will be adhered to.
- The management team, for a project or practice, should be broadly competent, if planning and monitoring systems are to function well. Much of this competence would relate to the management of people and the stresses they experience, for example, managers should give full recognition to the successes

of their staff, while allowing them to admit their inadequacies without fear of losing respect.

- All levels of staff must be prepared to quickly draw management attention to problems and support one another in resolving them.
- People who are planning work at different levels of management should communicate about their concerns and what they are doing about them.
- Communication between the various design offices must be effective and broad. This applies not only to design information and work planning, but also to the rationale that underlies design decisions and the organisation of work.

APPENDIX: AN EXAMPLE OF A DESIGN WORK PLAN

This Appendix provides an example of a design work plan that might be seen in practice. It is not discussed in the text of the book, where diagrams are provided to clarify particular points. It is included to broaden the reader's appreciation of the examples given in the text, so that the detailed advice can be seen in relation to the kind of co-ordinating plan that is most likely to be encountered in a real project. This example is developed from an actual project carried out in the 1980s. It may not be in full accordance with current practice and readers are invited to take a critical view of it, rather than adopting it as a model. In particular, the specialist contractors' programme of design work is not included here, whereas this should be developed together with the programme of manufacture and site operations.

The set of planning charts suggests the amount of detail that is likely to be included in the top levels of a work programme, for a building and project of typical complexity. The actual amount of detail required would depend to some extent on the type of project, the complexity of the design, whether or not the design is innovative and the confidence of the design team to co-ordinate their work with one another in the given time-scale. However, additional detail would generally shown on sub-charts, rather than the top-level charts shown here. Where there is less confidence that work can be done within the available time, work planning should be done in greater detail.

The first chart in the Appendix sets out the principal stages of the project as a series of stages or sub-projects. This master plan contains bars that represent the sub-projects, each of which is developed in more detail on the succeeding sub-charts. In practice, a number of the bars shown on the sub-charts would also be analysed in detail. The hammocking facility, provided by most network planning computer software, would facilitate this. Lines that link activities can be confusing, if it is not obvious where they begin and end. This is most likely to happen where the timings of activities overlap and the arrows on the Gantt chart cross one another and other activities which they are not dependent on. There can be additional confusion in design work, where links can be made at the middle of a bar. It is therefore normal to include only the key dependencies, omitting those that are less critical, otherwise the charts become cluttered and the sequence of priority tasks will not stand out as it should.

Note that the links between sub-charts appear as short arrows that do not connect with bars on the same sub-chart. The term *rods* on the Manufacture and Site Operations sub-chart refers to full-size drawings of components that must be specially made.

In design work, planning should not be regarded as a prescriptive device, but as an integrative activity, which helps the members of the design team to co-ordinate their actions. Staff time for close planning and monitoring tends to be very restricted, throughout the design stages of a building project. Consequently, decisions must be made about what details should be incorporated in the plan. To compensate for the lack of detail in planning, and to ensure that the integrative function of planning is fulfilled, proactive communication should be fostered to assist the day-to-day co-ordination of work. Such communication has to be based on a full understanding of the brief, the acceptance criteria (see Section 2.4) and the design itself, as it develops. This illustrates the broad dependency of planning, and success in working to a plan, on a systematic approach to managing the qualitative aspects of design work. A good balance has to be reached, in sharing managerial resources between work planning and monitoring and quality control activities.

The need to check how well the design output is co-ordinated can have a limiting effect on the opportunities to overlap the design and the construction phases. If construction is to proceed smoothly, design changes must be avoided to documentation that has been issued for the tendering of construction work. The points at which design information is frozen are shown on the charts. Preliminary design reviews are the last opportunity to revise the general arrangement of the building and engineering schematics. Critical design reviews offer the final opportunity to adjust the detail designs, prior to tendering and the production of cutting and ordering lists for construction. The detail design sub-chart in this example shows only three dates for critical design reviews, because it is best to cross-check the designs of a number of construction work packages for consistency at the same time, to minimise problems later on.

The procurement schedule, showing procedures for the selection of consultants and specialist contractors, could be planned on a separate chart. The detail of construction operations would also be shown on separate sheets, to avoid confusion with the design operations. Note, however, that the principal construction operations are shown in these charts, along with the design work, in order to establish the connections between these activities and their relevance to the timing and sequence of the design work programme.

These charts also omit the programmes for developing the project organisation and quality control system, which are of no concern to most members of the design team. The implications should, nevertheless, be considered when drawing up these top-level charts.

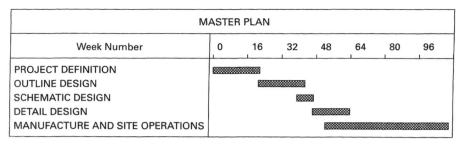

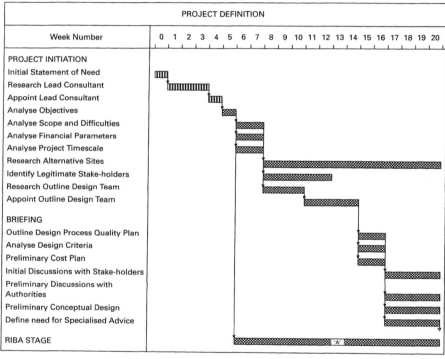

Bar Library 1

||||||| Client/Users   ▓▓▓▓ The entire design team

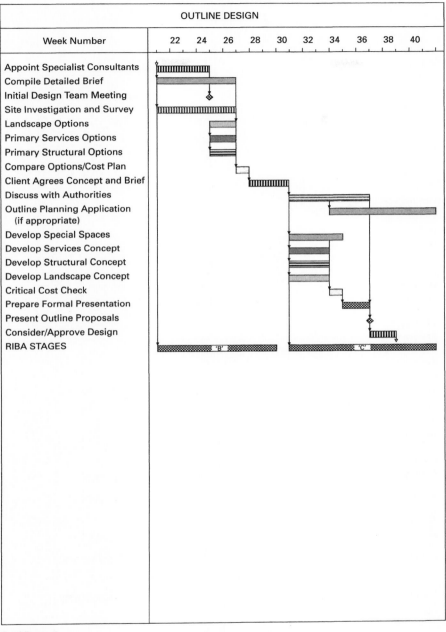

OUTLINE DESIGN

Week Number	22	24	26	28	30	32	34	36	38	40

Appoint Specialist Consultants
Compile Detailed Brief
Initial Design Team Meeting
Site Investigation and Survey
Landscape Options
Primary Services Options
Primary Structural Options
Compare Options/Cost Plan
Client Agrees Concept and Brief
Discuss with Authorities
Outline Planning Application
(if appropriate)
Develop Special Spaces
Develop Services Concept
Develop Structural Concept
Develop Landscape Concept
Critical Cost Check
Prepare Formal Presentation
Present Outline Proposals
Consider/Approve Design
RIBA STAGES

Bar Library 1

IIIIIII Client/Users   Architect   IIIIIII Surveyor
Landscape Arch.   Services Eng.   Structural Eng.
Cost Consultant   Authorities

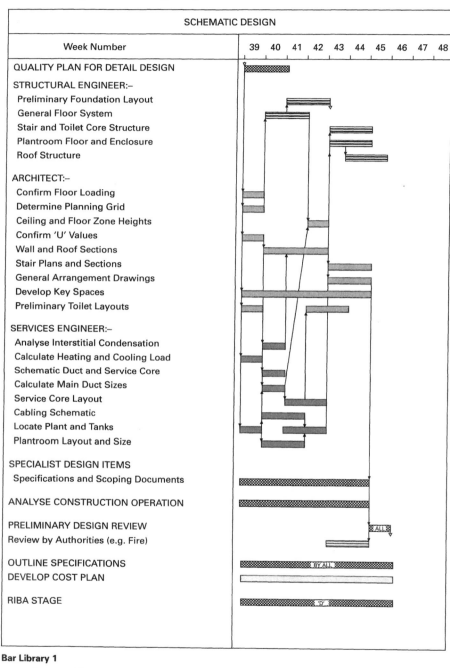

SCHEMATIC DESIGN

Week Number	39 40 41 42 43 44 45 46 47 48

QUALITY PLAN FOR DETAIL DESIGN

STRUCTURAL ENGINEER:–
 Preliminary Foundation Layout
 General Floor System
 Stair and Toilet Core Structure
 Plantroom Floor and Enclosure
 Roof Structure

ARCHITECT:–
 Confirm Floor Loading
 Determine Planning Grid
 Ceiling and Floor Zone Heights
 Confirm 'U' Values
 Wall and Roof Sections
 Stair Plans and Sections
 General Arrangement Drawings
 Develop Key Spaces
 Preliminary Toilet Layouts

SERVICES ENGINEER:–
 Analyse Interstitial Condensation
 Calculate Heating and Cooling Load
 Schematic Duct and Service Core
 Calculate Main Duct Sizes
 Service Core Layout
 Cabling Schematic
 Locate Plant and Tanks
 Plantroom Layout and Size

SPECIALIST DESIGN ITEMS
 Specifications and Scoping Documents

ANALYSE CONSTRUCTION OPERATION

PRELIMINARY DESIGN REVIEW
Review by Authorities (e.g. Fire)

OUTLINE SPECIFICATIONS
DEVELOP COST PLAN

RIBA STAGE

Bar Library 1

⬛ Structural Eng.   ⬛ Architect   ⬛ Services Eng.
⬛ Authorities   ☐ Cost Consultant

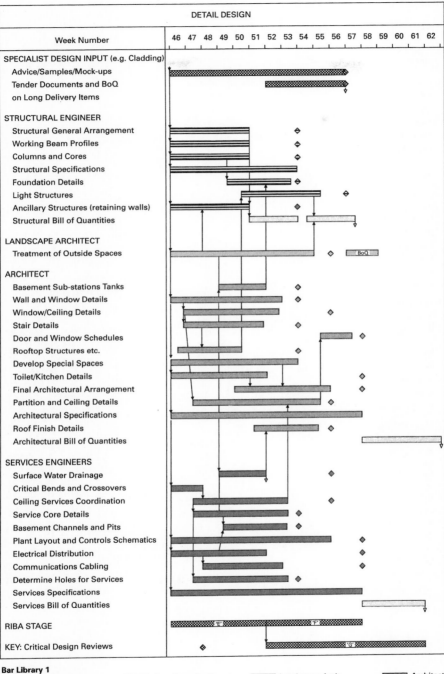

DETAIL DESIGN	

Week Number: 46 47 48 49 50 51 52 53 54 55 56 57 58 59 60 61 62

SPECIALIST DESIGN INPUT (e.g. Cladding)
- Advice/Samples/Mock-ups
- Tender Documents and BoQ
- on Long Delivery Items

STRUCTURAL ENGINEER
- Structural General Arrangement
- Working Beam Profiles
- Columns and Cores
- Structural Specifications
- Foundation Details
- Light Structures
- Ancillary Structures (retaining walls)
- Structural Bill of Quantities

LANDSCAPE ARCHITECT
- Treatment of Outside Spaces

ARCHITECT
- Basement Sub-stations Tanks
- Wall and Window Details
- Window/Ceiling Details
- Stair Details
- Door and Window Schedules
- Rooftop Structures etc.
- Develop Special Spaces
- Toilet/Kitchen Details
- Final Architectural Arrangement
- Partition and Ceiling Details
- Architectural Specifications
- Roof Finish Details
- Architectural Bill of Quantities

SERVICES ENGINEERS
- Surface Water Drainage
- Critical Bends and Crossovers
- Ceiling Services Coordination
- Service Core Details
- Basement Channels and Pits
- Plant Layout and Controls Schematics
- Electrical Distribution
- Communications Cabling
- Determine Holes for Services
- Services Specifications
- Services Bill of Quantities

RIBA STAGE

KEY: Critical Design Reviews

Bar Library 1

- Structural Eng.
- Services Eng.
- Cost Consultant
- Landscape Arch.
- Architect

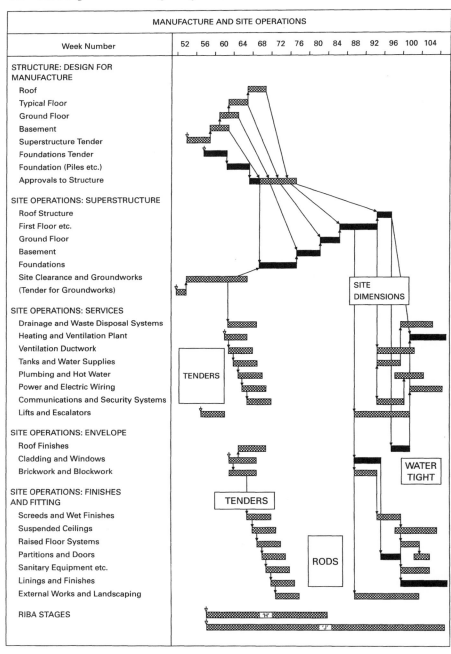

Note: The design professions and specialisms are not indicated on this particular sub–chart. Activities shown in solid black are on a critical path.

BIBLIOGRAPHY

The following list extends beyond the references made in the text, to include works that contributed to earlier drafts.

Association of Consultant Architects 1988 *Project Team Guidelines*.

Association of Project Managers 1988 *Proceedings of Internet World Congress of Project Management*, Glasgow.

Beheshti, M. R. (Ed.) 1985 *Proceedings of the International Design Participation Conference*. Eindhoven Technical University.

Bond van Nederlandse Architekten 1984 *Informatie-ververking Tijdens het Bouwproces*.

Bond van Nederlandse Architekten 1995 *Kwaliteitszorg voor Architecten*, 2nd edition.

Brighton, R. 1992 How to manage your time effectively. *Training Officer*, July/August.

British Standards Institute *Use of Network Techniques in Project Management, BS 6046:* 1981 Parts 2 and 4; 1984 Part 4; 1992 Part 3.

The Chartered Institute of Building 1992 *Code of Practice for Project Management for Construction and Development*.

Churchill, W. S. 1959 *The Second World War* (Abridged version). Cassell and Co. Ltd. First published 1948.

Coles, E. J. 1990 Occasional Paper No. 40: Design management: a study of practice in the building industry. The Chartered Institute of Building.

Co-ordinating Committee for Project Information 1987 *A Common Arrangement of Work Section for Building Works*. NBS Services Ltd. (Copyright ACE, BEC, RIBA, RICS.)

Davis; Cochrane 1987 Optimisation of future manpower requirements in a multi-discipline consultancy. *Construction Management and Economics, Vol. 5*, pp. 45–56.

Day, A.; Faulkener, A.; Happold, E. 1986 Computers and the organisation of design. *Proceedings of the IABSE Workshop*, Zurich, May.

Duggar, J. F. 1984 *Checking and Co-ordinating Architectural and Engineering Working Drawings*. McGraw Hill.

Fellows, R.; Langford, D.; Newcombe, R.; Urry, S. 1983 *Construction Management in Practice*. Longman Scientific and Technical.

Gray, C.; Coles, E. J. 1988 The identification of information transfer between specialists to form a design process model – report on Science and Engineering Research Council (grant D/31034). University of Reading.

Gray, C.; Flannagan, R. 1984 US productivity starts on the drawing board. *Construction Management and Economics*, Autumn, pp. 133–34.

Gray, C.; Hughes, W.; Bennet, J. 1994 *The Successful Management of Design*. University of Reading.

Harris, F.; McCaffer, R. 1983 *Modern Construction Management*. Granada. First published 1977.

Heller, J. J. 1992 *Catch 22*. Jonathan Cape. First published, Corgi Books, 1964.

Hall, P. 1980 *Great Planning Disasters*. Weidenfield and Nicholson.

Harrison, F. L. 1992 *Advanced Project Management: A Structured Approach*. Third edition. Gower.

Jones, J. C. 1981 *Design Methods – Seeds of Human Futures*. Wiley Interscience.

Kast, F.; Rozenswieg. J. 1981 *Organisation and Management – a Systems and Contingency Approach* (International Student Edition). McGraw Hill.

Lock, D. 1992 *Project Management*. Gower.

Lucas, J. 1987 Guide to architectural business management (series). *Architect's Journal*, October and November.

National Economic Development Council 1987 *Achieving Quality on Building Sites*.

Royal Institute of British Architects 1965, reprinted 1983, *Plan of Work for Design Team Operation*.

Topalian, A. 1979 *Management of Design Projects*. Associated Business Press.

Turner, J. R.; Speiser, A. 1992 Programme management and its information systems requirements. *International Journal of Project Management, Vol. 10, No. 4*, November.

Vietsch, C. A. 1987 Anamnese, diagnose, therepie – een onderzoek naar de bouwvoorbereiding van algemene ziekenhuizen. PhD thesis. Eindhoven Technical University.

Walker, A. 1984 *Project Management in Construction*. Granada.

Webb, A. 1992 The origin and use of cost performance measurement: part 2. *Project Manager Today*, January.

Webb, A. 1992 Configuration management. *Project Manager Today*, July/August.

INDEX

Printed and bound by CPI Group (UK) Ltd, Croydon, CR0 4YY
02/05/2025
01859319-0001